NELSON VICscience

Philipa Andrieux

Aaron Woolnough

SKILLS WORKBOOK

biology

VCE UNITS ③ + ④

Nelson VICscience Biology Units 3 & 4 Skills Workbook
1st Edition
Philipa Andrieux
Aaron Woolnough
Julie Gould
ISBN 9780170452618

Publisher: Eleanor Gregory
Editor: Felicity Clissold
Cover design: Leigh Ashforth (Watershed Art & Design)
Text design: Alba Design
Project designer: James Steer
Permissions researcher: Debbie Gallagher
Typeset by: SPi Global
Production controller: Renee Tome

Any URLs contained in this publication were checked for currency during the production process. Note, however, that the publisher cannot vouch for the ongoing currency of URLs.

Acknowledgements
Scenario 1, 2 & 3 p.12, 13, 14: Reproduced from scenarios created by the Murdoch Children's Research Institute (MCRI) for the Victorian Genethics Competition 1999-2005 with the permission of MCRI, the Gene Technology Access Centre and the Victorian Department of Education and Training.
Fig. 9.7 p.189: data source: https://bioone.org/journals/Copeia/volume-2003/issue-1/0045-8511(2003)003[0034:VAAIIA]2.0.CO;2/Variableand-Asymmetric-Introgression-in-a-Hybrid-Zone-inthe/10.1643/0045-8511(2003)003[0034:VAAIIA]2.0.CO;2.short
Fig.10.15 p.216: Based on material provided by the National Health and Medical Research Council, https://www.nhmrc.gov.au/about-us/publications/nationalstatement-ethical-conduct-human-research-2007-updated-2018, CC BY 3.0 Au

For product information and technology assistance,
in Australia call **1300 790 853**;
in New Zealand call **0800 449 725**

For permission to use material from this text or product, please email
aust.permissions@cengage.com

ISBN 978 0 17 045261 8

Cengage Learning Australia
Level 7, 80 Dorcas Street
South Melbourne, Victoria Australia 3205

Cengage Learning New Zealand
Unit 4B Rosedale Office Park
331 Rosedale Road, Albany, North Shore 0632, NZ

For learning solutions, visit **cengage.com.au**

Printed in Singapore by C.O.S. Printers Pte Ltd.
2 3 4 5 6 7 25 24 23 22 21

Contents

Introduction

Biology is the study of living organisms, their structure, function, growth, evolution, where they live and how they live. Like the study of any other science there is key knowledge and terminology that you need to know and understand and be able to use appropriately. However, no study of Biology would be complete without also addressing the key science skills. The study of Biology is not just about learning content, it is also about developing, using and demonstrating the skills that enable you to fully understand, experience and engage with the subject. It is about learning to think and work like a scientist.

Some of the key science skills can only be used and demonstrated in a laboratory situation but many can be developed, used and demonstrated away from the laboratory, and it is these skills that are the focus of this workbook.

Seven key science skills have been mandated by the Victorian Curriculum Assessment Authority (VCAA) across all VCE science subjects. These key science skills are transferable across subjects as well as being examinable in the VCE exam. Developing these key science skills means that you will be able to do the following.

- Develop aims and questions, formulate hypotheses and make predictions
- Plan and conduct investigations
- Comply with safety and ethical guidelines
- Generate, collate and record data
- Analyse and evaluate data and investigation methods
- Construct evidence-based arguments and draw conclusions
- Analyse, evaluate and communicate scientific ideas

(VCAA VCE Biology Study Design 2022–2026 pages 7–9)

Each of the key science skills listed above is broken up into multiple sub-skills. The mapping provided on pages vii–xi of this workbook allows you to see how these skills and sub-skills have been addressed in this workbook.

This workbook follows the structure of the *VCE Biology Study Design 2022–2026*. It is full of activities that have been carefully crafted to enable you to consolidate your knowledge on a topic and to develop, use and demonstrate key science skills. Developing any skill takes time and practice; the key science skills in this book have been introduced in a graduated way starting with **practising** skills that have been introduced to in previous years of science study. As you gain proficiency and confidence, you will go on to **reinforce** newer and more complex skills. Then there are the new skills requiring an increased level of proficiency and thinking that you will **develop** during the course.

Activities often require a mixture of skill levels for completion. Each activity has been signposted with an icon to indicate the highest level of skill required to complete the activity.

 Shows you activities that require previously introduced skills and will require practise as you work through the activities.

 Shows activities that will build on previously introduced skills.

 Shows activities that introduce a new skill or skills that require development and challenge you at a high level of proficiency.

This workbook can be used with any VCE Biology textbook that covers the *VCAA VCE Biology Study Design 2022–2026*. It has been mapped to *VICscience Biology Units 3 & 4* using icons in both the workbook and the student textbook. The icons have been placed in the workbook notifying you of the pages in the student textbook where corresponding content occurs, and conversely icons have been placed in the student textbook to indicate the best place to undertake each activity.

The major headings in the workbook match the major headings in *VICscience Biology Units 3 & 4*. Applicable key knowledge is listed under each of the major headings. Skill activities show the corresponding key skills listed that students will be using or demonstrating.

Acknowledgement

Many people have been involved in the production of this workbook, and special thanks go to:

Aunty Zeta Thompson is a respected Elder and descendant of the Yarra Yarra Clan of the Wurundjeri people on her paternal side and a descendant of the Ulupna Clan of the Yorta Yorta people on her maternal side. Aunty Zeta provided invaluable information and anecdotes about Aboriginal culture and life that formed the basis of several activities in this workbook.

Rebecca Farmlonga is a proud Wadawurrung woman and Traditional Owner. Rebecca has taught and led in secondary schools for more than 20 years and is passionate about Aboriginal and Torres Strait Islander Education. Rebecca reviewed the activities relating to Aboriginal and Torres Strait Islander peoples.

Julie Gould teaches VCE Biology at Brunswick Secondary College. Julie reviewed all the activities in the workbook and wrote the answers that appear in the back of this book.

> **Enjoy your study of VCE Biology and take the time to develop, use and demonstrate the key science skills that are an integral part of this course.**

Key science skill grid

Key science skill	VCE Biology Units 1–4	Chapters										
		1	**2**	**3**	**4**	**5**	**6**	**7**	**8**	**9**	**10**	**11**
Develop aims and questions, formulate hypotheses and make predictions Practise	• identify, research and construct aims and questions for investigation	1.1.2		3.3	4.1.2 4.3.2 4.3.3		6.2.2					
	• identify independent, dependent and controlled variables in controlled experiments	1.1.2			4.3.1 4.3.2 4.3.3	5.1.3 5.1.4 5.1.5						
	• formulate hypotheses to focus investigation	1.1.2			4.1.2 4.3.1 4.3.2 4.3.3			7.4		9.3.1		
	• predict possible outcomes	1.1.2	2.3.2		4.1.2 4.3.1 4.3.3	5.1.3 5.1.4 5.1.5		7.4		9.3.1		
Plan and conduct investigations Practise	• determine appropriate investigation methodology: case study; classification and identification; controlled experiment; correlational study; fieldwork; literature review; modelling; product, process or system development; simulation	1.1.3			4.3.1	5.1.3						
	• design and conduct investigations; select and use methods appropriate to the investigation, including consideration of sampling technique and size, equipment and procedures, taking into account potential sources of error and uncertainty; determine the type and amount of qualitative and/or quantitative data to be generated or collated	1.1.4 1.1.6				5.1.3	6.1.2					
	• work independently and collaboratively as appropriate and within identified research constraints, adapting or extending processes as required and recording such modifications						6.1.2					

Key science skill	VCE Biology Units 1–4	Chapters										
		1	2	3	4	5	6	7	8	9	10	11
Comply with safety and ethical guidelines Reinforce	• demonstrate safe laboratory practices when planning and conducting investigations by using risk assessments that are informed by safety data sheets (SDS), and accounting for risks	1.1.4				5.1.3	6.1.2					
	• apply relevant occupational health and safety guidelines while undertaking practical investigations							7.1.3				
	• demonstrate ethical conduct when undertaking and reporting investigations							7.1.3				
Generate, collate and record data Reinforce	• systematically generate and record primary data, and collate secondary data, appropriate to the investigation, including use of databases and reputable online data sources			3.6							10.4.1	
	• record and summarise both qualitative and quantitative data, including use of a logbook as an authentication of generated or collated data				4.3.2		6.1.2					
	• organise and present data in useful and meaningful ways, including schematic diagrams, flow charts, tables, bar charts and line graphs			3.5	4.1.2 4.3.2		6.3.1	7.1.3 7.5.1			10.1.1 10.5.2	11.1.1 11.2.3 11.3.1 11.4.1
	• plot graphs involving two variables that show linear and non-linear relationships	1.2.1							8.1.1			

Key science skill	VCE Biology Units 1–4	Chapters										
		1	2	3	4	5	6	7	8	9	10	11
Analyse and evaluate data and investigation methods Develop	• process quantitative data using appropriate mathematical relationships and units, including calculations of ratios, percentages, percentage change and mean	1.2.1				5.1.4						
	• identify and analyse experimental data qualitatively, handing where appropriate concepts of: accuracy, precision, repeatability, reproducibility and validity of measurements; errors (random and systematic); and certainty in data, including effects of sample size in obtaining reliable data	1.1.5 1.1.6				5.1.5	6.5					
	• identify outliers, and contradictory or provisional data									9.6.1		
	• repeat experiments to ensure findings are robust					5.1.3	6.1.2					
	• evaluate investigation methods and possible sources of personal errors/mistakes or bias, and suggest improvements to increase accuracy and precision, and to reduce the likelihood of errors			3.6								

Key science skill	VCE Biology Units 1–4	Chapters										
		1	2	3	4	5	6	7	8	9	10	11
Construct evidence-based arguments and draw conclusions Develop	• distinguish between opinion, anecdote and evidence, and scientific and non-scientific ideas			3.8		5.3.1						11.2.2
	• evaluate data to determine the degree to which the evidence supports the aim of the investigation, and make recommendations, as appropriate, for modifying or extending the investigation								8.3.1			
	• evaluate data to determine the degree to which the evidence supports or refutes the initial prediction or hypothesis				4.3.2	5.1.3 5.1.5			8.3.1	9.6.1	10.2.1	
	• use reasoning to construct scientific arguments, and to draw and justify conclusions consistent with the evidence and relevant to the question under investigation				4.3.2	5.1.3 5.1.4 5.1.5			8.1.1	9.3.2	10.1.2 10.1.3 10.3.1 10.3.2 10.4.1 10.5.1	11.1.1 11.2.3 11.4.2
	• identify, describe and explain the limitations of conclusions, including identification of further evidence required					5.1.3 5.1.5						11.2.2 11.3.1 11.4.3
	• discuss the implications of research findings and proposals					5.1.4	6.2.2		8.4.2		10.3.1	11.2.2

9780170452618

Key science skill	VCE Biology Units 1–4	Chapters										
		1	2	3	4	5	6	7	8	9	10	11
Analyse, evaluate and communicate scientific ideas Develop	• use appropriate biological terminology, representations and conventions, including standard abbreviations, graphing conventions and units of measurement		2.6.2							9.6.2		
	• discuss relevant biological information, ideas, concepts theories and models and the connections between them					5.2.3 5.2.4		7.1.1 7.1.2 7.3.2	8.3.1	9.4.2		
	• analyse and explain how models and theories are used to organise and understand observed phenomena and concepts related to biology, identifying limitations of selected models/theories		2.1.2 2.2.1 2.2.4 2.4	3.4 3.7.3	4.2.1							
	• critically evaluate and interpret a range of scientific and media texts (including journal articles, mass media communications and opinions in the public domain), processes, claims and conclusions related to biology by considering the quality of available evidence		2.2.5		4.1.3				8.2 8.4.2			11.4.3
	• analyse and evaluate bioethical issues using relevant approaches to bioethics and ethical concepts, including the influence of social, economic, legal and political factors relevant to the selected issue	1.1.7		3.3		5.3.1			8.3.2	9.4.3 9.5.1		
	• use clear, coherent and concise expression to communicate to specific audiences and for specific purposes in appropriate scientific genres, including scientific reports and posters	1.3	2.5			5.2.1 5.2.2			8.1.1 8.1.2 8.3.2	9.3.2		
	• acknowledge sources of information and assistance, and use standard scientific referencing conventions								8.1.1 8.4.2			

1 Designing and conducting a scientific investigation

Remember

TB
PAGE 4

In Units 1 and 2 of VCE Biology, you learnt about the scientific method. This knowledge will help you work though this chapter of the workbook. Test yourself by answering the following questions from memory or complete this section to consolidate your knowledge as you work through the chapter.

1 What do scientists test using the scientific method?

_____

2 Jane wrote the following hypothesis in her science logbook: 'If the plant is kept in light, then it will grow taller.'

   a Identify the independent variable.

_____

   b Identify the dependent variable.

_____

   c What are three other variables that would need to be controlled in this investigation?

_____

_____

_____

   d Predict the outcome of the investigation if the hypothesis is supported.

_____

_____

3 When designing a scientific investigation, state two important factors that need to be considered.

_____

_____

1.1 Investigation design

Key knowledge

Investigation design

- biological concepts specific to the selected scientific investigation and their significance, including definitions of key terms
- characteristics of the selected scientific methodology and method, and appropriateness of the use of independent, dependent and controlled variables in the selected scientific investigation
- techniques of primary quantitative data generation relevant to the selected scientific investigation
- the accuracy, precision, reproducibility, repeatability and validity of measurements
- the health, safety and ethical guidelines relevant to the selected scientific investigation

Each of the following tasks will help you identify and articulate the different components of the scientific method. As you work through the tasks, questions and readings, make sure you think about how the various parts of the scientific method interact.

1.1.1 Observation

Consolidation of knowledge   PAGE 5

When you dig in the garden, you might notice an abundance of organisms in the soil. You can see some organisms, such as ants, mites and springtails, but others you cannot see, such as bacteria, fungi and spores. You can sometimes smell fungal spores when you disturb them with a spade.

Earthworms are usually very obvious in garden soil, especially moist soil. They vary in size and can be as large as your little finger in circumference. Sometimes when digging in the garden, you may cut an earthworm in half. You may notice that in parts of the garden where you have not been digging, you do not see any earthworms.

Figure 1.1 Digging in the garden

1   What is an observation? What senses can you use to make an observation?

2   What are two observations a Biology student might record from the above scenario?

3   State the criteria for an effective research question.

4   Choose one of the observations from your answer to Question 2 and use it to write a research question. Make sure you meet all the requirements for an effective research question.

5   Using the research question from Question 4, circle the independent variable in red and the dependent variable in blue. Rewrite your research question if necessary.

6   Define 'hypothesis'.

7   State the criteria for an effective hypothesis.

8   Turn your research question from Question 5 into a hypothesis, ensuring that you meet all the requirements for an effective hypothesis.

1.1.2 Designing an investigation

Key science skills
Develop aims and questions, formulate hypotheses and make predictions
- identify, research and construct aims and questions for investigation
- identify independent, dependent and controlled variables in controlled experiments
- formulate hypotheses to focus investigation
- predict possible outcomes

Practise
PAGE 8

Clare and Isobel had seen advertisements for Charlie Carp™, the liquid organic fertiliser made from fish. Clare and Isobel are keen gardeners and wanted to find out which concentration of Charlie Carp™ would make tomato plants produce the most fruit. They purchased seven tomato plants and placed them on a table in front of some windows, as shown in Figure 1.2.

1 Write an aim for the investigation.

_____

_____

_____

2 Write a research question for the investigation.

_____

Figure 1.2 Clare and Isobel's set-up for their investigation

_____

_____

3 Turn the research question into a hypothesis that can be tested by a scientific investigation.

_____

_____

4 Identify the independent variable.

_____

5 Identify the dependent variable.

_____

6 Identify any extraneous variables and explain how they would be controlled.

_____

_____

_____

_____

_____

7   What is the purpose of the control plant (labelled 1 in Figure 1.2)?

8   State the conditions of the control plant (labelled 1 in Figure 1.2).

9   What results would support the hypothesis?

10   What results would refute the hypothesis?

1.1.3 Methodologies

Key science skills

Plan and conduct investigations
- determine appropriate investigation methodology: case study; classification and identification; controlled experiment; correlational study; fieldwork; literature review; modelling; product, process or system development; simulation

Practise

The methodology is the broad framework of the approach taken in the investigation. The choice of methodology depends on the aim of the investigation.

Complete Table 1.1 by writing in the best methodology for each investigation aim. Choose from the following methodologies.

- » Case study
- » Product, process or system development
- » Fieldwork
- » Literature review
- » Classification and identification

- » Modelling
- » Correlational study
- » Simulation
- » Controlled experiment

Table 1.1 Research methodologies and investigation aims

Methodology	Investigation aim
	To determine if the type of pasture affects the density of grasshoppers
	To determine if different coloured light affects the reproduction rate of bacterial cells
	To investigate the historical impact of the redirection of the waterflow on the Merri Creek ecosystem
	To investigate the relationship between land clearing and breeding numbers of orange-bellied parrots in a western Tasmanian forest
	To develop an artificial hand to assist people with arthritis
	To use a digital model of a bacterial cell to study the effect of an antibiotic on the cell wall
	To determine how many different types of organisms live in my garden
	To review the current literature on the genomes of lions from the African nation of Ghana
	To determine the mathematical relationship between the distribution and density of antelope in African grasslands

1.1.4 Designing your investigation to test your hypothesis

Key science skills
Plan and conduct investigations
- design and conduct investigations; select and use methods appropriate to the investigation, including consideration of sampling technique and size, equipment and procedures, taking into account potential sources of error and uncertainty; determine the type and amount of qualitative and/or quantitative data to be generated or collated

Comply with safety and ethical guidelines
- demonstrate safe laboratory practices when planning and conducting investigations by using risk assessments that are informed by safety data sheets (SDS), and accounting for risks

Reinforce

TB
PAGE 11

1 Select the best methodology from Table 1.1 to test the aim of Clare and Isobel's investigation (Figure 1.2 on p. 5). Justify your choice of methodology.

2 For Clare and Isobel's investigation determine:

a the data that Clare and Isobel would be collecting during their investigation.

b whether the data is qualitative and/or quantitative.

c how Clare and Isobel can ensure that their data is:
i valid
ii precise
iii accurate

d how often data would be collected and the likely duration of the investigation.

e what equipment would be needed to collect the data.

f what units the quantitative data would be measured in.

g any potential sources of error or uncertainty in the data.

3   Construct a materials list for Clare and Isobel's investigation. Make sure that you include every item that they will need to use during their investigation and how many of each item will be required.

_____

_____

_____

_____

_____

_____

_____

_____

_____

4   Use Table 1.2 to construct a risk assessment for Clare and Isobel's investigation.

Table 1.2 Risk assessment

What are the risks in doing this investigation?	How can these risks be managed to stay safe?

5   Construct a method for Clare and Isobel's investigation that considers the materials being used and the data being collected. Check back on your materials list and ensure that all items required are accounted for in the method.

_____

_____

_____

_____

_____

_____

_____

_____

_____

_____

1.1.5 Primary data

Sarita plays basketball and wants to ensure that she is the best at three-point shots in her school. Her coach tells her that she needs to be precise with her shooting, while her brother tells her that she needs to be accurate. Sarita is very confused by these two statements and cannot distinguish between the two. Sarita then remembers that her Biology teacher had used these terms when explaining scientific investigations. Her Biology teacher had drawn four concentric circles to help explain the difference between accuracy and precision (Figure 1.3). Sarita recalls that accuracy refers to how close the measured value is to the true value whereas precision refers to how close repeated measured values are to each other.

1 Figure 1.3a shows low accuracy and low precision. Complete Figure 1.3b–d by adding in seven black dots to represent the caption below each set of circles.

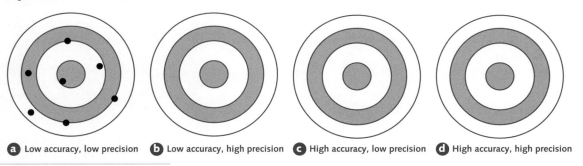

a Low accuracy, low precision   **b** Low accuracy, high precision   **c** High accuracy, low precision   **d** High accuracy, high precision

Figure 1.3 Accuracy and precision

2 Three groups of students set up an experiment to investigate the effect of glucose concentration on the rate of fermentation in yeast. The rate of fermentation was to be measured by the temperature change in the fermenting mixture. Before the students started their experiments, they were asked by their teacher to test their thermometers to make sure they were working properly. The students recorded their results in Table 1.3. Decide how accurate and precise each set of results is. Use rating words such as 'high', 'low', 'moderate', 'very' and 'not very'.

Table 1.3 Thermometer readings

Group	Thermometer reading (°C)			Electronic thermometer reading (°C)	Accuracy and precision of each group's results	
	Measurement 1	Measurement 2	Measurement 3		Accuracy	Precision
1	20.0	19.0	19.5	22.1		
2	23.0	23.0	22.5	21.9		
3	20.0	21.0	23.0	21.1		

3 What advice would you give to the students to improve the:

a accuracy of their results?

_____

_____

b precision of their results?

_____

_____

1.1.6 Minimising error

PAGE 14

Key science skills
Analyse and evaluate data and investigation methods
- identify and analyse experimental data qualitatively, handling where appropriate concepts of: accuracy, precision, repeatability, reproducibility and validity of measurements; errors (random and systematic); and certainty in data, including effects of sample size in obtaining reliable data

Develop

While you are planning and conducting your investigation, you need to be careful that you do not introduce error into your method or data collection. There are four main types of error to be aware of: personal error, systematic error, random error and bias.

1   Define each type of error and state how you could overcome each error in a scientific investigation.

   a   Personal error

   _____

   _____

   _____

   b   Systematic error

   _____

   _____

   _____

   c   Random error

   _____

   _____

   _____

   d   Bias

   _____

   _____

   _____

2   What might indicate to you that an error has occurred in your scientific investigation?

   _____

   _____

   _____

3   For each of the following situations, state which of the four main types of error is involved.

   a   James and Louise are working together on their biology project. They are investigating whether different coloured lights affect the growth of the plant *Pelargonium australe*. To collect data, they measure the mass of each of the 10 plants every day at 1 pm. They use biology lab scales that have not been calibrated correctly, which results in each measurement starting at 10 g, rather than from zero.

   _____

b Amir and May are investigating the effects of exercise on young people. They choose 10 students from their class. They place five of the students into Group A and ask them to exercise for an extra 30 minutes every day for one month. They place the other five students into Group B and ask them to continue their normal exercise routine. Group A contains five members of the school cross-country team, whereas Group B contains no members of a school sports team.

c Terry is testing whether the aquatic microorganism *Planaria* is affected by different concentrations of salt. He has set up 10 beakers with different salt concentrations and is counting the number of *Planaria* in each beaker over a two-week period. He accidently knocks one of the beakers and about 40 mL of water (which may or may not have contained *Planaria*) spills onto the bench. He tops up the beaker with tap water and continues with his investigation.

d Chen has set up her experimental apparatus to investigate the hypothesis that the more caffeine a woman consumes, the higher her heart rate. Chen commences with 10 women of similar ages in the same room over a 30-minute period. The caffeine solution runs out when it has been administered to six participants. Another caffeine solution is mixed and administered to the remaining four participants.

1.1.7 Ethical guidelines

Key science skills
Analyse, evaluate and communicate scientific ideas
- analyse and evaluate bioethical issues using relevant approaches to bioethics and ethical concepts, including the influence of social, economic, legal and political factors relevant to the selected issue

TB
PAGE 17

Develop

Ethics are moral principles that govern a person's behaviour. Ethics is knowing the difference between right and wrong. Why is an understanding of ethics relevant to the Biology classroom? Many scientific investigations use plants, fungi and micro-organisms, all of which are considered non-sentient beings. However, scientists also carry out research using animals, including humans. For example, in the quest to find a vaccine for the coronavirus, SARS-CoV-2, during the 2020 epidemic, scientists used ferrets as a model for humans. Ferret lungs react in a similar fashion to human lungs on exposure to the virus.

Before undertaking the research, the researchers would have had to put their case to use ferrets to an ethics committee. They would have used one of the three approaches shown in the ethical dilemma triangle in Figure 1.4. They most likely argued that it was a consequence-based approach, and that the sacrifice of a small number of ferrets could save millions of human lives.

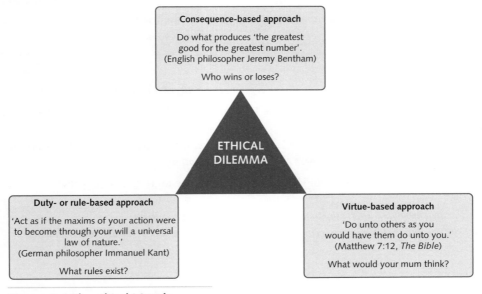

Figure 1.4 The ethical triangle

Bioethics refers to ethical issues in a biological context. When you are considering bioethical issues, you can also consider the following concepts. These concepts can be used on their own or in conjunction with an ethical approach.

i **Integrity**: the commitment to searching for knowledge and understanding and the honest reporting of all sources of information and communication of results, whether favourable or unfavourable, in ways that permit scrutiny and contribute to public knowledge and understanding.

ii **Justice**: the moral obligation to fairly consider competing claims; that there is no unfair burden on a particular group from an action; and that there is fair distribution and access to the benefits of an action.

iii **Beneficence**: the commitment to maximising benefits and minimising the risks and harms involved in taking a particular position or course of action.

iv **Non-maleficence**: involves avoiding the causations of harm. However, as positions or courses of actions in scientific research may involve some degree of harm, the concept of non-maleficence implies that the harm resulting from any position or course of action should not be disproportionate to the benefits from any position or course of action.

v **Respect**: involves consideration of the extent to which living things have an intrinsic value or instrumental value; giving due regard to the welfare, liberty and autonomy, beliefs, perceptions, customs and cultural heritage of both the individual and the collective; consideration of the capacity of living things to make their own decisions; and when living things have diminished capacity to make their own decisions, ensuring that they are empowered where possible and protected as necessary.

Based on material provided by the National Health and Medical Research Council, https://www.nhmrc.gov.au/about-us/publications/national-statement-ethical-conduct-human-research-2007-updated-2018, CC BY 3.0 Au

When considering an ethical dilemma, use the following as a guide.

Step 1 Identify the ethical dilemma.

Step 2 Identify which of the three approaches in Figure 1.4 (p. 11) is the driving force behind the ethical dilemma.

Step 3 Decide which of the five concepts above could also apply; you can use more than one.

Read Scenario 1 and consider your responses to the three steps provided above. To help you articulate these, an example has been completed for you. Take some time to think about your own responses and those that have been provided. Do your responses align with the example provided?

Once you feel comfortable with answering these questions using the example as a guide, complete Scenarios 2 and 3, using your own reasoning. Make sure you support your reasoning with a sound rationale.

Scenario 1: Genetic screening

A mutation in a gene that causes deafness has recently been identified. A number of employers in the airline industry are concerned about this finding and want to screen prospective new employees for carrier status of the D (deafness) allele. They will refuse to employ people as pilots or flight attendants who might be adversely affected by the pressure changes associated with flying because they are concerned about the safety of their employees and customers, and possible litigation issues.

Step 1 Identify the ethical dilemma.
Should employers be allowed to genetically screen potential employees and use this as a basis to award a job?

Step 2 Identify which of the three approaches in Figure 1.4 is the driving force behind the ethical dilemma. Provide reason(s) for your choice.
Consequences-based – the employers are placing importance on screening out people with the deafness allele in an effort to keep their employees and customers safe and avoid litigation.

Step 3 Decide which of the five concepts could also apply; you can use more than one. Provide reason(s) for your choice.
Beneficence – the employers are trying to avoid harm being caused to others.
Non-maleficence – the employers are trying to avoid employing someone who might cause harm.

Scenario 2: Gene fix

GENE Dreams, the utopian biotechnology company that brought you the 'BABY BODY microchip', has just announced the release of a 'new and even better' technology to ensure the health of your newborn child.

Previously, GENE Dreams could diagnose genetic conditions and parents could discard the embryo if they didn't want a baby with that condition. Now, GENE Dreams has perfected gene therapy techniques that will allow parents to correct gene mutations detected by their BABY BODY microchip.

If untreated, the mutations could cause deafness, short stature or a predisposition to breast or colon cancer in the child. Instead of an abortion, 'BABY BODY gene fix' will allow a quick and easy correction of the mutation. The DNA microarray BABY BODY microchip would screen embryos generated by IVF techniques even before being implanted into the mother's womb. If a gene mutation is found, an adenovirus would be used to insert a correct copy of the mutant gene into the embryo. Every cell in the baby's body will contain the new gene and it will be passed on to their children, ensuring healthy children for generations to come.

GENE Dreams claims that this technology is very safe and has had a success rate of 99% in their trials. However, a leaked document has shown that of the 50 babies born through this technology, two have developed a rare genetic condition called retinoblastoma. GENE Dreams commented vigorously that 'It is the right of every parent to choose whether or not to have a baby with or without a disability and to desire the best possible health outcomes for their children'.

A spokeswoman commented that up to 4% of babies from normal conceptions are also born with a disability, and that there was no need for alarm due to the two reported cases of retinoblastoma. She said that they have developed the technology to ensure choice for parents willing to pay the modest price of $20 000 for each gene therapy treatment, and that it should appeal particularly to religious groups opposing abortion, since it provides treatment rather than discarding embryos.

Step 1   Identify the ethical dilemma.

_____

Step 2   Identify which of the three approaches in Figure 1.4 is the driving force behind the ethical dilemma. Provide reason(s) for your choice.

_____

Step 3   Decide which of the five concepts could also apply; you can use more than one. Provide reason(s) for your choice.

_____

_____

_____

Scenario 3: A heartfelt quandary

Grace has experienced much tragedy in her life. Eighteen months ago, her 34-year-old brother died in his sleep, and just six months ago a young cousin collapsed and died at a rock concert. Grace has very few memories of her father, who died suddenly when she was just five years old. All three deaths were caused by heart attacks. Grace's medical and family history indicates that she also might be at risk of hypertrophic cardiomyopathy.

After much deliberation, she decided to undergo genetic testing. The test found that she has a mutation in the MYH7 gene associated with hypertrophic cardiomyopathy. While she has taken her doctor's advice to modify her lifestyle, including avoiding competitive sports and strenuous exercise, she plans to keep up her job as the local school bus driver.

'There's no way I'm telling my boss about this,' Grace said to her doctor. 'My husband left me recently after his business went bankrupt and it's up to me to support my two kids. My boss would have me out the door in a flash if he thought there was something wrong with me, and then what would I do? I don't know how to do anything else. I'm taking care of myself, I'm a vegetarian and I do yoga twice a week. That's more than I can say for some of the other drivers. They're fat and smoke like chimneys; they'll drop dead before me.'

Mr Moral Quandary works as the receptionist for Grace's doctor. He is a member of school council, and his two children aged 14 and 16 travel on the local school bus. He is aware of Grace's test results and plans to inform school council of Grace's genetic risk status.

'It's a question of the broader interests of the community being more important than Grace's right to privacy. My children travel on that bus and I would be remiss in my duties if I ignored the fact that she is endangering our community's most precious assets,' he said.

Mr Quandary's plans have enraged his wife, Justice Quandary, who is a prominent civil rights activist.

'Grace will suffer discrimination if you reveal her genetic status to the school committee,' she argued. 'I want our children to remain on the bus to demonstrate the importance of civil freedom. Plus, how do you know what the test results mean? Grace may be at no more risk of dropping dead than you or me.'

Step 1   Identify the ethical dilemma.

_____

_____

Step 2   Identify which of the three approaches in Figure 1.4 (p. 11) is the driving force behind the ethical dilemma. Provide one or more reasons for your choice.

_____

_____

Step 3   Decide which of the five concepts could also apply; you can use more than one. Provide one or more reasons for your choice.

_____

_____

_____

_____

_____

1.2 Scientific evidence

Key knowledge

Scientific evidence
- the nature of evidence that supports or refutes a hypothesis, model or theory
- ways of organising, analysing and evaluating primary data to identify patterns and relationships, including sources of error and uncertainty
- authentication of generated primary data through the use of a logbook
- assumptions and limitations of investigation methodology and/or data generation and/or analysis methods

1.2.1 Analysing your data

Key science skills

Generate, collate and record data
- plot graphs involving two variables that show linear or non-linear relationships

Analyse and evaluate data and investigation methods
- process quantitative data using appropriate mathematical relationships and units, including calculations of ratios, percentages, percentage change and mean

Develop

TB
PAGE 19

When plants photosynthesise, they take in carbon dioxide and produce oxygen as a by-product of the reaction. Dane and Brett, two Year 12 Biology students, assumed that the rate of photosynthesis could be measured by the amount of oxygen a plant produced. A high amount of oxygen produced means that the plant is photosynthesising rapidly. They designed an investigation in which they measured the rate of photosynthesis by counting the number of oxygen bubbles produced in a 10-minute period under different light intensities. Dane and Brett set up six samples of pondweed inside six upturned funnels. Test tubes were fitted to the stems of the funnels to catch the oxygen bubbles. These were placed inside beakers that were filled with distilled water and sodium hydrogen carbonate (a source of carbon dioxide) and the beakers were put in front of a lamp at different distances (Figure 1.5). Dane and Brett performed three trials at each distance to get three sets of data for each distance.

Dane and Brett counted the number of bubbles produced in a 10-minute period and recorded the data in Table 1.4. The data in the table is raw (unanalysed) data. You will need to analyse the data to find out what it is telling you. Look for trends and relationships within the data, which is best done through graphing. Answer the questions below, which will assist you in analysing the data.

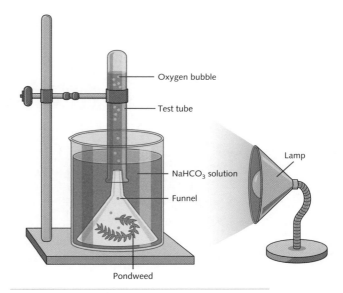

Figure 1.5 The equipment set-up

Table 1.4 The number of oxygen bubbles produced at various distances from the lamp

Pondweed sample	Distance from lamp (cm)	Number of oxygen bubbles produced in 10 minutes			
		Trial 1	Trial 2	Trial 3	Mean
1	10	30	36	34	
2	20	14	24	12	
3	30	6	8	6	
4	40	2	4	5	
5	50	1	1	3	
6	60	0	1	0	

Part A

1   Formulate a hypothesis for Dane and Brett's investigation.

_____

_____

2   What is the independent variable?

_____

3   What is the dependent variable?

_____

4   Calculate the mean number of oxygen bubbles for each distance from the lamp. Record these in the last column of Table 1.4 (p. 15).

5   Dane wanted to draw a graph to show the relationship between light intensity and the number of bubbles produced. He said that the graph should be a bar graph. Brett disagreed.

   a   Which type of graph would best represent the data? Explain your answer.

_____

_____

   b   The grid below is a sample of the graph paper used in your logbook. Plot the data from Table 1.4 using the correct type of graph. Ensure you follow the guidelines for correctly drawing a scientific graph.

c Describe the relationship shown in the graph.

_____

_____

d Using your graph, predict the number of bubbles that would be produced if the lamp is 25 cm from the pondweed.

_____

6 Brett argued that they should have set up a control. What would this control consist of and why does Brett think it is important?

_____

_____

_____

Part B

Questions 7 and 8 will extend your learning. You will engage more deeply through a more detailed written discussion and conclusion in response to this investigation.

7 Write a discussion for this investigation of no more than 200 words. In your discussion, you need to include:

- a summary of the results and what this means with respect to relevant biological concepts
- a statement of how the results significantly add to the collective scientific knowledge
- any limitations of the investigation and the reasons why
- suggestions for future research.

Your discussion must be well thought out, succinct and based on what was discovered during the investigation. Do not:

- repeat the results
- include statements of analysis that the results do not show
- introduce new data
- use too much jargon.

_____

_____

_____

_____

_____

_____

_____

8 Write a conclusion of no more than 60 words. In your conclusion you need to:

- restate the hypothesis
- state whether the hypothesis was supported or refuted
- state the main findings of the investigation
- recommend what to investigate next as a result of your findings.

_____

_____

_____

_____

1.3 Science communication

Key knowledge
Science communication
- conventions of science communication: scientific terminology and representations, symbols, formulas, standard abbreviations and units of measurement
- conventions of scientific poster presentation including succinct communication of the selected scientific investigation, and acknowledgements and references
- the key findings and implications of the selected scientific investigation.

1.3.1 Presenting your work as a scientific poster

Key science skills
Analyse, evaluate and communicate scientific ideas
- use clear, coherent and concise expression to communicate to specific audiences and for specific purposes in appropriate scientific genres, including scientific reports and posters

Develop

1 Unit 4, Outcome 3 requires you to demonstrate communication skills by presenting your research findings in the form of a scientific poster. Your logbook forms part of the assessment for Unit 4, Outcome 3, so make sure that your logbook includes the details shown in Table 1.5.

Table 1.5 Monitor your logbook requirements

✓	Element to be included in your logbook
	Investigation planning
	Identification and management of relevant risks
	Recording of raw data
	Preliminary analysis of results
	Evaluation of results
	Identification of outliers and how you treated them

2 The word limit for the poster is 600 words. Calculate the word limits for each section of the poster by completing Table 1.6. Make sure that your total number of words is not more than 600.

Table 1.6 Word counts for poster sections

Poster section	Maximum number of words
Title	
Introduction	
Methodology and method	
Results	
Discussion	
Conclusion	
References	N/A
Acknowledgements	N/A
Total number of words	**600**

9780170452618

3 The poster may be produced in hard copy or electronically. Either way, you need to follow VCAA specifications on how to set out your poster.

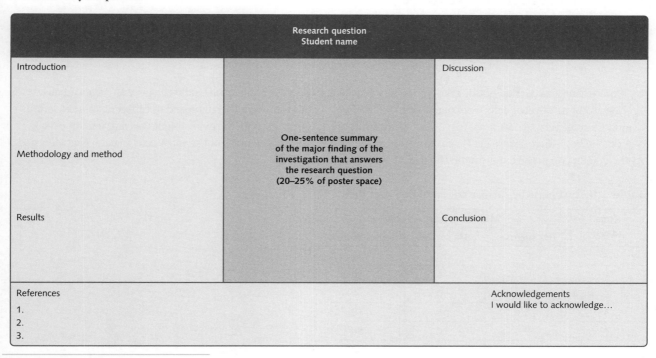

Figure 1.6 A poster template

1.4   Chapter review

1.4.1 Key terms

Use different coloured pens, pencils or highlighters to match the key terms to their definitions.

Table 1.7 Key terms and definitions

Key term	Definition
Accuracy	How similar two series of experimental results are to each other when conducted by different scientists
Dependent variable	The broader framework of approach taken in a scientific investigation to test a hypothesis
Hypothesis	The comparison of how close a measurement is to the true value
Independent variable	The steps taken to carry out a scientific investigation
Method	The variable that is measured during an investigation
Methodology	The variable that is manipulated during an investigation
Outlier	Errors stemming from the scientist conducting the experiment
Personal errors	How close a series of measurements are to each other, without reference to true value
Precision	Errors stemming from measurements being off from true value consistently
Quantitative data	How useful the collected data is for showing the link between the dependent and the independent variables
Qualitative data	A value given for a measurement if it was measured under perfect conditions
Repeatability	A prediction based on limited evidence that can be expanded upon after further experimentation
Reproducibility	Errors affecting precision that arise from unpredictable variations in the measurements
Random errors	When results are affected only by a single independent variable
Systematic errors	Experimental results where observational characteristics are provided
True value	a range of values that the true value falls within
Uncertainty	A measurement that is vastly different from the rest of the data in the experiment
Validity	Experimental results where numerical values of measurements are provided

Chapter review (continued)

1.4.2 Exam practice

Exam
practice

The following information relates to Questions 1–3.

An experiment was carried out by students to test the effect of sucrose concentration on the growth of yeast cells. Precise amounts of yeast cells were placed into sterile Petri dishes that were then exposed to different concentrations of glucose solutions: 0%, 5% and 10%. All other variables were kept constant. The experiment was carried out over 5 days. The Petri dishes were observed every day at the same time and the mass of the yeast cells in each was recorded. At the conclusion of the experiment, the results were recorded in Table 1.8.

Table 1.8 Yeast mass by sucrose concentration over time

Time (days)	Mass (mg) of yeast cells		
	0% sucrose	5% sucrose	10% sucrose
0	0	0	0
1	0	5	20
2	0	10	40
3	0	15	60
4	0	20	80
5	0	25	100

1 ©VCAA 2019 Q7–9 (adapted) EASY Which one of the following hypotheses is supported by the results?
   A If the yeast cells grow for 4 days, then the Petri dishes will be completely covered in yeast cells.
   B If the concentration of sucrose increases, then the yeast cells will grow more quickly.
   C If the yeast cells are kept in the dark, then the yeast cells will grow more slowly.
   D If the yeast cells grow faster, then the temperature of the location will increase.

2 In this experiment, the independent variable is ©VCAA MEDIUM
   A time.
   B concentration of sucrose.
   C the number of yeast cells.
   D the mass of yeast cells in the Petri dish.

3 The students wanted to check the validity of their data. The students should ©VCAA EASY
   A repeat the experiment several times to find out if they would obtain the same data.
   B organise their data into a different format to help identify a trend.
   C change the independent variable in the experiment.
   D make sure no extraneous variables are affecting the dependent variable.

The following information relates to Questions 4 and 5.

An experiment was conducted to test the following hypothesis about the effect of the plant growth hormone indoleacetic acid:

Concentrations of indoleacetic acid above 0.01 parts per million stimulate shoot growth but inhibit root growth.
In the experiment, radish seedlings were grown in different concentrations of indoleacetic acid, as shown in Table 1.9.

Table 1.9 Radish seedling shoot and root growth in indoleacetic acid

Concentration of indoleacetic acid (parts per million)	Stimulation (+) or inhibition (−) of shoot growth (%)	Stimulation (+) or inhibition (−) of root growth (%)
0	0	0
0.00001	+0.10	−30
0.0001	+6	−50
0.001	−20	−70
0.01	−60	−85
1	−70	−90
10	−80	−95
100	−90	−100

4 ©VCAA 2016 Q15 (adapted) EASY Which one of the following is a reasonable conclusion to draw from the results of the experiment?

A The data is not accurate enough to be able to draw a conclusion.
B The hypothesis is supported.
C The hypothesis is refuted.
D The hypothesis is proven.

5 In this experiment, the dependent variable is

A concentration of indoleacetic acid.
B radish seedlings.
C time.
D percentage change in shoot and root growth.

6 ©VCAA 2016 Q22 (adapted) EASY *Staphylococcus aureus* is a common cause of food poisoning. Data was collected and analysed for the occurrence of illness caused by this organism in Queensland over a 5-year period.

Figure 1.7 shows the average monthly notifications per 100 000 people for the illness caused by *S. aureus*.

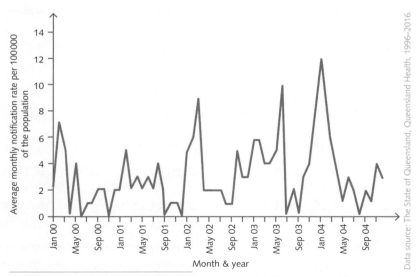

Data source: The State of Queensland, Queensland Health, 1996–2016.

Figure 1.7 Average monthly notification rate for illness caused by *Staphylococcus aureus*

It can be concluded from the data that

A there are three periods during which the notification rate was greater than eight per 100 000.
B the notification rate was always lowest during January of each year.
C the notification rate over a 5-year period showed a steady trend.
D the notification rate in 2002 was highest in August.

2 The relationship between nucleic acids and proteins

Getty Images Plus/E+/alanphillips

Remember

PAGE 38

In Unit 1 of VCE Biology, you learnt about cell structure and cell division. This knowledge will help you work though this chapter of the workbook. Read through the information below and test yourself by answering the questions that follow from memory or complete this section to consolidate your knowledge as you work through the chapter.

Deoxyribonucleic acid (DNA) is found in the nuclei of eukaryotic cells where it is packaged into chromosomes (Figure 2.1). Humans have 46 chromosomes in each of their body cells (and 23 in each of their gametes).

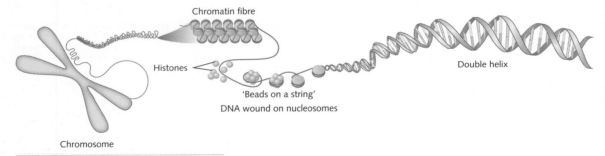

Figure 2.1 DNA is wound tightly and packaged into chromosomes

Chromosomes are replicated during the two processes of nuclear division.

- Mitosis (for asexual reproduction, growth, repair and replacement of cells) results in identical daughter cells (chromosome number = 2n).
- Meiosis (for the production of gametes for sexual reproduction) results in cells that contain half the usual number of chromosomes (chromosome number = n).

Offspring inherit DNA from their parents through sperm (n) and ova (n), which after fertilisation form the zygote (2n) – the first cell of the new individual. DNA carries the genetic code from one generation to the next. This chapter explores the components that make up DNA molecules, looks at how DNA carries the genetic code for the production of proteins, and explains the steps that occur inside our cells that lead to proteins being produced.

Proteins are important in cells because all enzymes are proteins. Enzymes determine which chemical reactions occur in cells, which molecules are produced and which molecules are broken down. Without these chemical reactions, there would be no life.

Using your knowledge of biology and the information you have just read, answer the following questions.

1   What does DNA carry the instructions to produce?

_____

2   What are the monomers that make up proteins?

_____

3   List two functions of proteins in cells.

_____

_____

4   Where would you find the following structures inside a cell?

   a   DNA

_____

   b   Endoplasmic reticulum

_____

   c   Ribosomes

_____

   d   Golgi apparatus

_____

2.1   Nucleic acids

Key knowledge

The relationship between nucleic acids and proteins
- nucleic acids as information molecules that encode instructions for the synthesis of proteins: the structure of DNA, the three main forms of RNA (mRNA, rRNA and tRNA) and a comparison of their respective nucleotides

2.1.1 Structure of nucleic acids

Consolidation of knowledge   TB PAGE 40

DNA is a nucleic acid. DNA is a double-stranded molecule that is twisted around itself in the shape of a double helix (Figure 2.2). If you could zoom in on a DNA molecule with a very powerful microscope, you would see that it is made up of smaller units or monomers. These monomers are called nucleotides. A DNA molecule is made up of a large number of nucleotides joined together to make the two long strands of DNA that are twisted together.

1   Part of Figure 2.2 is a line drawing representing a nucleotide. Using colour can assist recall. Use coloured pencils to colour in the nucleotide, colouring the phosphate (yellow), sugar (green) and base (orange). Label each component as you colour it in.

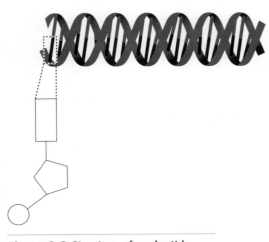

Figure 2.2 Structure of nucleotide

Now that you have identified the structure of a nucleotide, answer the following questions to explore nucleotides in greater depth.

2   List the four types of bases in the nucleotides that make up DNA. What letters are used to represent each of these bases?

3   Which part of the nucleotide holds the two strands of DNA together?

4   What type of bond holds the two strands of DNA together?

5   Which parts of the nucleotide form the backbone of each strand of DNA?

6   What type of bond holds the nucleotide backbone of DNA together?

2.1.2 Structure of DNA: building a model

TB
PAGE 41

Key science skills
Analyse, evaluate and communicate scientific ideas
• analyse and explain how models and theories are used to organise and understand observed phenomena and concepts related to biology, identifying limitations of selected models/theories

Develop

James Watson and Francis Crick discovered the structure of DNA in 1953. They did this by building a model that accounted for all the scientific evidence about DNA that was known at the time. In this activity, you will use the template provided in Figure 2.3 on page 27 to build a model of an unwound DNA helix that shows the following attributes.

• Sugar–phosphate backbone
• Repeating nucleotide structure
• Antiparallel nature
• Complementary base pairing
• Non-template or coding strand with the sequence of bases: 5'ATCCGGTA3'
• Template or non-coding strand

After you have built your DNA model, answer the following questions.

1   What does DNA stand for? (Watch your spelling.)

2   **a**   Using the non-template or coding strand sequence on your DNA model write out the sequence of bases in triplets on the corresponding template or non-coding strand.

   b   What rule have you applied to do this?

3   One of the important attributes of DNA is that the distance between the two strands is approximately the same throughout the helix. Use a ruler to measure the distance between the two strands along your unwound helix and comment on:

   a   whether your model displays this attribute.

   b   the structural features of DNA that enable this to occur.

   c   how the complementary base pairing rule contributes to this attribute.

   d   Watson and Crick initially thought that the bases in DNA paired like with like. How would this affect this attribute of DNA?

4   The double helix is made up of two antiparallel strands (i.e. they run in opposite directions). How does your model demonstrate this feature?

5   Describe the key features of your model.

6   All scientific models have limitations in what they can show. Describe the limitations of your model (i.e. those features of the real DNA molecule that the model cannot show correctly or at all).

7   State one way that you could improve your model.

_____

_____

8   State two concepts about DNA that you now understand better as a result of building your model.

_____

_____

_____

_____

_____

Activity: build a model of DNA

In this activity, you will create a model of unwound DNA showing eight bases on its non-template or coding strand: 5'ATCCGGTA3'. You will need scissors and a glue stick.

What to do

Step 1 Cut out the individual nucleotides below (Figure 2.3).

Step 2 Glue the nucleotides onto the section labelled **DNA** in Figure 2.4 (p. 32) to produce an unwound DNA molecule.

Step 3 Start at the top of the page building the 5' strand first and then add the 3' strand.

Step 4 Review the attributes of DNA that you need to show from page 24. Label these attributes on your model.

Step 5 You will add the complementary RNA strand to this model when you complete section 2.2.1.

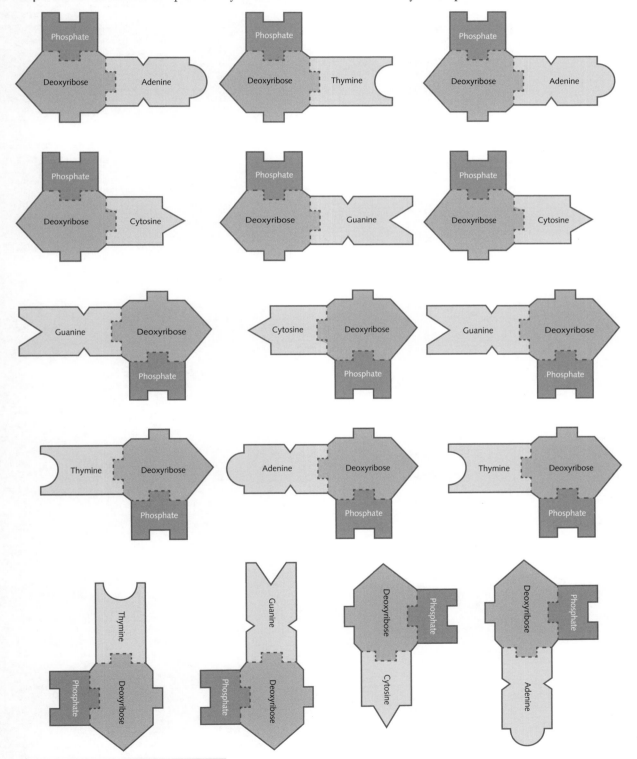

Figure 2.3 Individual nucleotides to use to build your own model of DNA

2.1.3 Structure of RNA

RNA is another type of nucleic acid. Like DNA it is made of nucleotide monomers, but there are some very important differences in both structure and function.

Consolidation of knowledge   PAGE 45

Use Table 2.1 to show how DNA is similar to and different from RNA.

Table 2.1 Features of DNA and RNA

	DNA molecule	RNA molecule
Is it made up of nucleotides?		
What is the type of sugar in its sugar–phosphate backbone?		
What are its nitrogen bases?		
Where is it located in eukaryotic cells?		
Is it a stable molecule?		
How many strands in each molecule?		

1   Using information from Table 2.1, compare DNA and RNA. When you compare, you state how the two molecules are similar and different. Make sure you mention both DNA and RNA for each feature that you are comparing.

_____

_____

_____

_____

_____

_____

_____

_____

2   There are many different types of RNA. VCE Biology Units 3 and 4 focuses on three of these RNA molecules and their roles in protein synthesis.

Complete Table 2.2 to show the functions of these three types of RNA.

Table 2.2 Functions of three types of RNA

	mRNA	tRNA	rRNA
What is its full name?			
Where is it located in the cell?			
What does it do?			
How does its name give you a clue about what it does?			
What is its shape?			

2.2   Gene expression

2.2.1 Transcription

Develop

DNA is found in the nuclei of cells. DNA carries the genetic code for the production of proteins, which are made by ribosomes in the cytosol of cells. DNA molecules are too large to fit through the pores of the nuclear membrane to get to the ribosomes, so a smaller intermediate molecule carries information from the DNA to the ribosomes. This intermediate molecule is called messenger RNA (mRNA).

The process of copying DNA to produce a complementary strand of mRNA molecule is called transcription. In this activity, you will model transcription by producing a pre-mRNA molecule for the template or non-coding strand of the DNA molecule you made in section 2.1.2.

After you have built your pre-mRNA model, answer the questions below.

1   Where in the cell does transcription occur?

2   State two functions of mRNA.

3   Refer to the sequence of bases on your DNA model (Figure 2.4).

    a   Write out the DNA triplets on the template or non-coding strand and the corresponding sequence of bases on the pre-mRNA strand.

    b   What do you notice about the sequence of bases on the non-template or coding strand of DNA and the pre-mRNA strand?

4   Explain why the non-template or coding strand of DNA has been given this name.

5 State the name of the enzyme responsible for transcription.

_____

6 State two differences between the DNA molecule and the pre-mRNA molecule.

_____

_____

_____

7 All scientific models have strengths in what they can show. Describe two features of transcription that are well represented by your model.

_____

_____

_____

_____

8 All scientific models have limitations in what they can show. Describe two features of transcription that are not well represented by your model.

_____

_____

_____

_____

9 Describe two ways in which you could improve your model.

_____

_____

_____

_____

10 State two concepts about transcription that you now understand better as a result of building this model.

_____

_____

_____

_____

Figure 2.4 Build your model DNA and pre-mRNA strands here.

5' DNA 3'	pre-mRNA

Activity: build a model of pre-mRNA

In Activity 2.1.2, you built a model of DNA. You will now return to this model to show transcription by adding the complementary pre-mRNA strand. You will need scissors and a glue stick.

What to do

Step 1 Identify the template or non-coding strand of your DNA model. (Hint: The template or non-coding strand runs from 3' to 5'.)

Step 2 Cut out all the individual nucleotides of RNA below (Figure 2.5).

Step 3 Use these bases to transcribe a pre-mRNA strand for the template or non-coding strand of DNA on page 32.

Step 4 Glue them onto the section of the page labelled **RNA** of page 32.

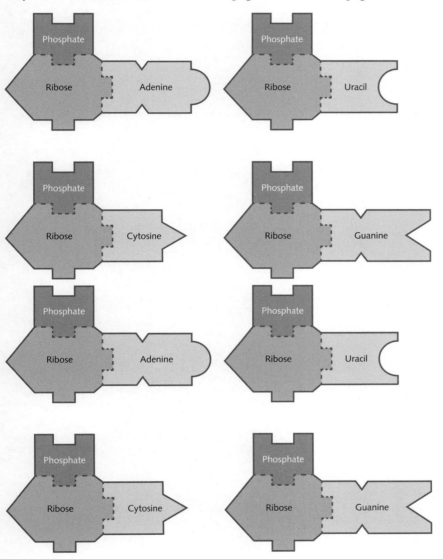

Figure 2.5 Individual nucleotides to use to build your own model of pre-mRNA

2.2.2 RNA processing

Check your RNA and transcription knowledge. Complete the gaps in the following statement.

Consolidation of knowledge **TB** PAGE 48

1   Transcription produces a linear molecule of pre-mRNA. Pre-mRNA contains non-coding regions called _____ and coding regions called _____.

   For this molecule to become mature mRNA, it undergoes RNA processing.

2   Figure 2.6 shows transcription in a cell nucleus. On Figure 2.6, label the DNA, pre-mRNA and the mature mRNA molecules.

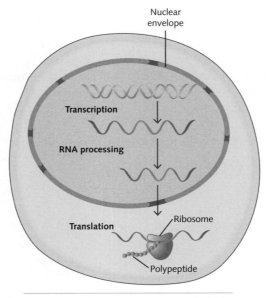

Figure 2.6 Transcription in a cell nucleus

3   State the names of two biological processes occurring in the nucleus of the cell in Figure 2.6.

   _____

   _____

4   Number in ascending order the following list of events occurring inside the nucleus of the cell shown in Figure 2.6.

   ☐ Mature mRNA ready to leave the nucleus

   ☐ RNA polymerase positioned onto the DNA

   ☐ Poly-A tail added 5′ methyl cap added

   ☐ DNA unwinds

   ☐ Introns removed

   ☐ Strand of pre-mRNA built

   ☐ Promoter region signals the start of a gene

   ☐ DNA unzips

   ☐ Exons spliced together

5   Write the mRNA sequences for the following triplets of **template** or non-coding DNA.

   a   CTG GAT TCC ATG

   _____

   b   TTC CGA GTC ACA

   _____

6   Write the mRNA sequences for the following triplets of non-template or **coding** DNA.

   a   ATC GCT ATC GTT

   _____

b TTA TAT CCT ATG GTG

7 Explain how one pre-mRNA molecule could form several different mature mRNA molecules. Use the name of this process in your explanation, and explain the biological consequence of this process.

8 Once the mRNA has been processed, it leaves the nucleus and carries the genetic code to organelles in the cytosol. State the name of these organelles.

9 Which type of RNA molecule is involved in the production of these organelles?

2.2.3 Translation

TB PAGE 50

On Figure 2.7, label the following features:

- Nucleus
- Plasma membrane
- Ribosome
- The process of transcription
- The process of translation

- mRNA
- tRNA
- Amino acids
- Growing polypeptide chain

Consolidation of knowledge

When you label a diagram, make sure that you:

- use a ruler to draw the lines
- do not use arrows for label lines
- ensure the label line points to the centre of the structure being labelled
- print all labels horizontally
- place all labels on the right-hand side of the diagram
- do not cross label lines.

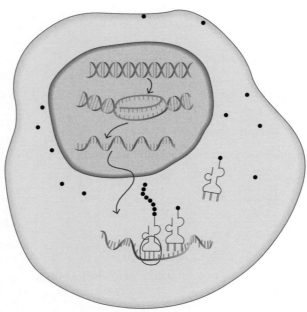

Figure 2.7 Understanding translation

2.2.4 The genetic code

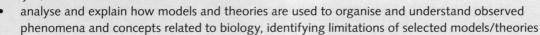

Key science skills

Analyse, evaluate and communicate scientific ideas

- analyse and explain how models and theories are used to organise and understand observed phenomena and concepts related to biology, identifying limitations of selected models/theories

Develop

TB

PAGE 51

A closer look at tRNA

Transfer RNA (tRNA) is a clover-leaf-shaped molecule (Figure 2.8). At the bottom of the clover leaf is a set of three bases called the anticodon. The anticodon binds to a complementary set of three bases on the mRNA molecule called the codon. The codon–anticodon relationship follows the complementary base pairing rule where A pairs with U and C pairs with G.

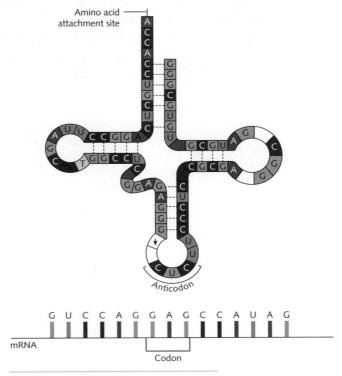

Figure 2.8 tRNA anticodon and mRNA codon

1   Determine the anticodons for each of the following codons.

   a   UUA _____

   b   ACG _____

   c   CCC _____

2   Determine the nucleotide sequence on the **template** or **non-coding DNA** strand for each of the following anticodons.

   a   AAU _____

   b   CGC _____

   c   GCG _____

3   Determine the nucleotide sequence on the **non-template** or **coding DNA** strand for each of the following anticodons.

   a   AAU _____

   b   CGC _____

   c   GCG _____

Activity: build a model of translation

In this activity, you will build a model of translation and protein synthesis. You will need scissors and a glue stick.

What to do

Step 1 In Figure 2.9 (p. 39), cut out the tRNA molecules individually, the mRNA molecule in one long strand and the amino acids individually.

Step 2 Use the genetic code in Figure 2.10 (p. 41) to determine which tRNA molecule matches with each amino acid. Remember that the genetic code uses the codon on the mRNA to code for amino acids (not the anticodon on the tRNA).

Step 3 Glue the mRNA strand halfway down this page. Match the anticodons on the tRNA with their associated amino acid molecules with the codons on the mRNA and glue into place.

Step 4 Draw peptide bonds between each amino acid to represent a polypeptide.

Step 5 Label your model to show: tRNA, mRNA, amino acids, peptide bonds and polypeptide.

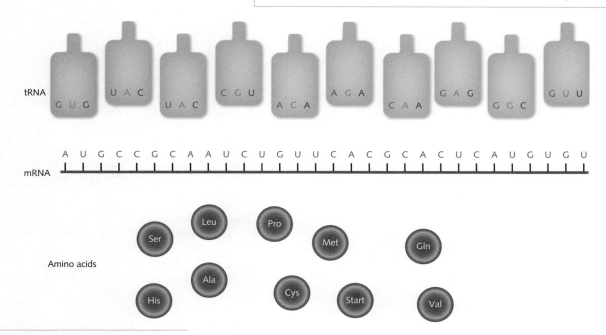

Figure 2.9 tRNA molecules, mRNA and amino acids

After you have built your model of translation and protein synthesis, answer the following questions.

1   Where in the cell are these processes occurring?

2   Does the mRNA use the series of bases known as the codon or the anticodon to decode the genetic code?

3   What is the significance of the AUG, UAA, UAG and UGA codons in the genetic code?

4   Figure 2.10 shows the genetic code.

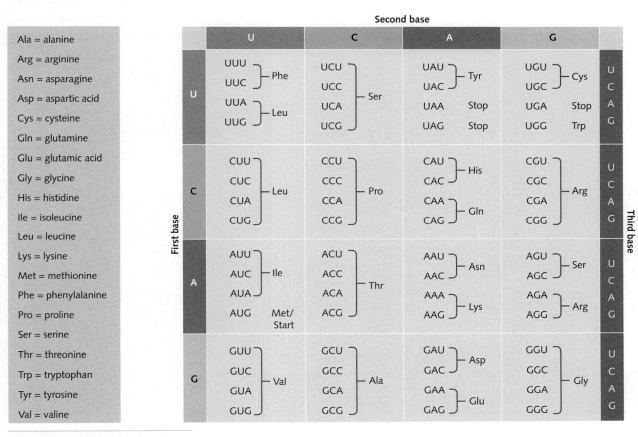

Figure 2.10 **The genetic code**

a   Sixty codons correspond to 20 amino acids. Use the correct terminology to explain the biological significance of this advantage.

b   State the sequence of amino acids in the final polypeptide chain that was produced in your model of translation.

c   Predict the effect on the final protein in your model if the following mistakes were made during translation.

   i   CAC instead of CAA

   _____

   _____

   ii   GCA instead of GCU

   _____

   _____

   iii UGA instead of AGA

   _____

   _____

5   All scientific models have limitations in what they can show. Evaluate your translation and protein synthesis model by comparing it with your knowledge of protein synthesis. Construct a table to show the strengths and limitations of your model.

6   Modify the instructions for this activity to improve at least one aspect of your model based on your evaluation.

   _____

   _____

   _____

   _____

7   Explain two concepts about translation and protein synthesis that you now understand better as a result of building this model.

   _____

   _____

   _____

   _____

2.2.5 Scientific literacy

Key science skills
Analyse, evaluate and communicate scientific ideas
- critically evaluate and interpret a range of scientific and media texts (including journal articles, mass media communications and opinions in the public domain), processes, claims and conclusions related to biology by considering the quality of available evidence

Develop

TB
PAGE 52

Carefully read the article below and answer the questions on page 44.

Squids can edit their RNA in an unprecedented way, scientists discover

(Jacinta Bowler, 25 March 2020)

When it comes to squids, you just can't keep them down.

Not just because they're slippery, but also because they have an incredible genetic editing ability – it lets them tweak their own RNA long after it's left the nucleus.

Here's what that means. Genes, in humans at least, mostly stay unchanging until they're recombined and passed onto the next generation.

This is the same for our messenger RNA (mRNA). Helpful molecules read our DNA, create short little RNA messages, and send them outside the nucleus to tell the rest of the cell which proteins need to be built.

Once that mRNA has exited the nucleus, it's thought the genetic information it carries can't be messed with much – but new research has shown that in squid nerves, this isn't the case.

'We are showing that squid can modify the RNAs out in the periphery of the cell', says Marine Biological Laboratory (MBL), Woods Hole geneticist Joshua Rosenthal.

'It works by this massive tweaking of its nervous system', Rosenthal told Wired. 'Which is a really novel way of going through life.'

The team took nerves from specimens of adult male longfin inshore squid (*Doryteuthis pealeii*), and analysed the protein expression, as well as the squids' transcriptome, which is similar to a genome, but for mRNA.

They found that in squid nerves (or neurons), the mRNA was being edited outside of the nucleus, in a part of the cell called the axon.

This mRNA editing allows the squids to finely tune the proteins they produce at local sites. With this finding, squids have become the only creatures we know of that can do this.

This isn't the first time squids have shown off their genetic editing prowess, though. Back in 2015,

a similar team at MBL discovered that squids edit their mRNA inside their nucleus to an incredibly large degree – orders of magnitude more than what happens in humans.

'We thought all the RNA editing happened in the nucleus, and then the modified messenger RNAs are exported out to the cell', Rosenthal explains.

But the team showed that although editing is happening in both, it occurs significantly more outside the nucleus in the axon, rather than inside the nucleus.

So, why do squids bother? Why do they need to change their mRNA so much? Well, we don't yet know, but the research team has some ideas.

Octopus, cuttlefish and squids all use mRNA editing to diversify the proteins produced in the nervous system. This could be one of the reasons why these creatures are so much smarter than other invertebrates.

'The idea that genetic information can be differentially edited within a cell is novel and extends our ideas about how a single blueprint of genetic information can give rise to spatial complexity', the team writes in their new paper.

'Such a process could fine-tune protein function to help meet the specific physiological demands of different cellular regions.'

Although right now this is just an interesting genetics study into squids, the researchers think that eventually, this type of system might be able to help treat neurological disorders that include axon dysfunction.

'RNA editing is a hell of a lot safer than DNA editing', Rosenthal told Wired.

'If you make a mistake, the RNA just turns over and goes away.'

'Squids Can Edit Their RNA in an Unprecedented Way, Scientists Discover', Jacinta Bowler, 25 March 2020, Science Alert

1   Do you think this article is providing reliable information? Give reasons for your answer.

2   Locate one section of the article that is written in everyday language and rewrite it with appropriate biological detail.

3   State the name of the process referred to in this statement: 'We thought all the RNA editing happened in the nucleus, and then the modified messenger RNAs are exported out to the cell …'

4   Explain the biology behind this statement: 'RNA editing is a hell of a lot safer than DNA editing. If you make a mistake, the RNA just turns over and goes away.'

5   Rewrite this passage so it could be understood by a 12-year-old child.

    Genes, in humans at least, mostly stay unchanging until they're recombined and passed onto the next generation. This is the same for our messenger RNA (mRNA). Helpful molecules read our DNA, create short little RNA messages, and send them outside the nucleus to tell the rest of the cell which proteins need to be built. Once that mRNA has exited the nucleus, it's thought the genetic information it carries can't be messed with much – but new research has shown that in squid nerves, this isn't the case.

6   Write two research questions that could lead to further research into this discovery.

7   Hypothesise about squids' ability to edit their mRNA before translation. Propose an investigation to test your hypothesis.

2.3   Gene regulation

Key knowledge
The relationship between nucleic acids and proteins
- the structure of genes: exons, introns and promoter and operator regions
- the basic elements of gene regulation: prokaryotic *trp* operon as a simplified example of a regulatory process

2.3.1 The *trp* operon: an example of gene regulation – Part A

Consolidation of knowledge   **TB** PAGE 55

Figure 2.11 shows two diagrams of *Escherichia coli* chromosomes, each specifically showing the *trp* operon. In Figure 2.11a, tryptophan is present in the cell's environment; in Figure 2.11b, tryptophan is not present in the cell's environment. Use these two diagrams to create a visual record along with a narrative to show your understanding of the functioning of the *trp* operon.

Make sure you include the following features where relevant either in your narrative or by adding features and labels to the diagrams.

- RNA polymerase
- Tryptophan
- Repressor protein

- mRNA
- Enzymes for tryptophan synthesis

Add to each diagram and use the space beside each diagram to write a narrative explaining what is occurring in each cell in relation to the *trp* operon.

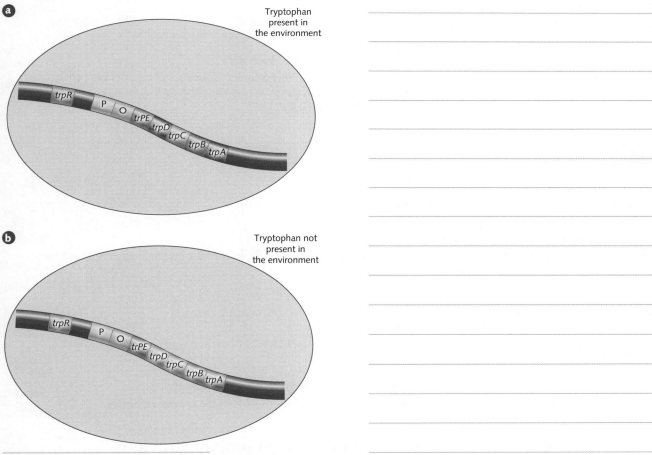

Figure 2.11 *Escherichia coli* **chromosomes showing** *trp* **operon (not drawn to scale): a tryptophan present in the environment and b tryptophan not present in the environment.**

46   UNIT 3 / AoS 1: What is the role of nucleic acids and proteins in maintaining life?

2.3.2 The *trp* operon: an example of gene regulation – Part B

Key science skills
Develop aims and questions, formulate hypotheses and make predictions
- predict possible outcomes

Practise

In the following experiment, two mutants of the *trp* operon are described.
- Mutant A is in the operator region of the *trp* operon and has a point mutation so that it is unable to bind with the repressor protein.
- Mutant B expresses a *trp* repressor protein that is unable to bind tryptophan.

For each statement a–j below select one alternative (A, B, C or D) that most closely matches the statement. Note: 'Wild type' means that it has a normal functioning *trp* operon.

A: Mutant A
B: Mutant B
C: Both A and B
D: Neither A nor B

a _____ In the presence of tryptophan, these cells behave like wild-type cells.
b _____ In the absence of tryptophan, these cells behave like wild-type cells.
c _____ These cells can synthesise tryptophan.
d _____ These cells can break down tryptophan.
e _____ Tryptophan can act as an inducer in these cells.
f _____ Tryptophan can act as a corepressor in these cells.
g _____ The structural genes of the *trp* operon can be transcribed in these cells.
h _____ In the presence of tryptophan, RNA polymerase binds to the promoter of the *trp* operon in these cells.
i _____ In the absence of tryptophan, RNA polymerase binds to the promoter of the *trp* operon in these cells.
j _____ These cells die if tryptophan is omitted from their medium.

2.4   Proteins

Key knowledge
The relationship between nucleic acids and proteins
- amino acids as the monomers of a polypeptide chain and the resultant hierarchical levels of structure that give rise to a functional protein
- proteins as a diverse group of molecules that collectively make an organism's proteome, including enzymes as catalysts in biochemical pathways

2.4.1 Proteins: building a model

Key science skills
Analyse, evaluate and communicate scientific ideas
- analyse and explain how models and theories are used to organise and understand observed phenomena and concepts related to biology, identifying limitations of selected models/theories

Develop

Activity: build a model to show the four levels of protein structure

In this activity, you will create a model of the four levels of protein structure. You will need scissors, a stapler, sticky tape and a glue stick.

What to do

Primary structure

Step 1   Cut out the first strip of amino acids from Figure 2.13 on page 49 by cutting down the vertical line. This is the primary structure of the protein – a chain of amino acids that has been produced by translation at the ribosome.

Secondary structure

Step 2   Cut out the rest of the strips from Figure 2.13 on page 49 by cutting down the vertical lines.

Step 3   Wind two of these strips around a pencil. Release them. Place them on the table so they become a spiral. Use the stapler to create bonds between the coils of each spiral (Figure 2.12a). This represents two alpha helices.

Step 4   Take two other strips and fold along the horizontal lines, turning the paper over with each fold. Form an arch with each strip and use the stapler to create bonds between the two sides of the arch (Figure 2.12b). This represents two pleated sheets.

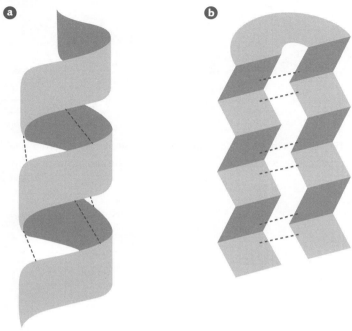

Figure 2.12 Secondary protein structure: a alpha helix; b pleated sheet

Tertiary structure

Step 5   Take the alpha helices and pleated sheets and use sticky tape to tape them end-on-end so that you have four strips sticky taped together to make one longer strip. Intermingle the longer strip so it forms a tangled ball. Use the stapler to create bonds to keep the tangled ball in place. This is the tertiary structure that forms polypeptide chains.

Step 6   Repeat steps 3–5 to make two more tertiary structures.

Quaternary structure

Step 7   Staple the three tertiary structures together to form a quaternary structure. This is the protein molecule.

After you have built your model of the four levels of protein structure, answer the questions below.

1   Where in the cell are these processes occurring?

_____

2   State the type of bond in the following levels of protein structure.

a   Primary level

_____

b   Secondary level

_____

3   What level of protein structure is found in enzymes? Explain why the shape of the enzyme molecule is important.

_____

_____

_____

4   Comment on how mistakes in translation could affect enzyme function. Infer the effect this could have on cell functioning.

_____

_____

_____

5   All scientific models have limitations in what they can show. Evaluate your protein model by comparing it with your existing knowledge of protein synthesis. Construct a table to show the strengths and limitations of your model.

6   Modify the instructions for this activity to improve at least one aspect of your model based on your evaluation.

_____

_____

7   Explain two concepts about protein structure that you now understand better as a result of building this model.

_____

_____

_____

_____

Figure 2.13 provides the strips required to build the four levels of protein structure. See pages 46–47 to find out what to do.

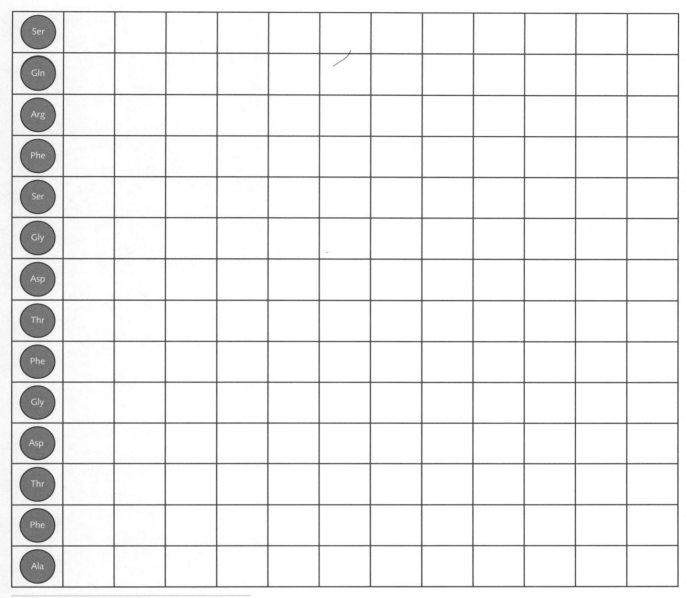

Figure 2.13 Strips to use to build protein molecule

2.5   The protein secretory pathway

Key knowledge
The relationship between nucleic acids and proteins
- the role of rough endoplasmic reticulum, Golgi apparatus and associated vesicles in the export of proteins from a cell via the protein secretory pathway.

2.5.1 Telling the story of the protein secretory pathway

Key science skills
Analyse, evaluate and communicate scientific ideas
- use clear, coherent and concise expression to communicate to specific audiences and for specific purposes in appropriate scientific genres, including scientific reports and posters

Develop

TB
PAGE 64

In order to communicate scientific ideas to others, you need to be able to use a combination of text and diagrams. This task has been designed to help you build this skill.

Step 1   On Figure 2.14, label the following in the space beside the diagram.

- DNA
- Pre-mRNA
- Mature mRNA
- Ribosome
- Rough endoplasmic reticulum
- Smooth endoplasmic reticulum

- Golgi apparatus
- Proteins
- Transport vesicle
- Secretory vesicle
- Plasma membrane

Step 2   Use the lined space on the right of Figure 2.14 to describe what is happening in the cell. Your story needs to be able to be understood by a Year 12 student who has not studied biology. You may draw arrows on the diagram to assist in understanding.

Use the following headings in your description: transcription, translation and protein secretory pathway.

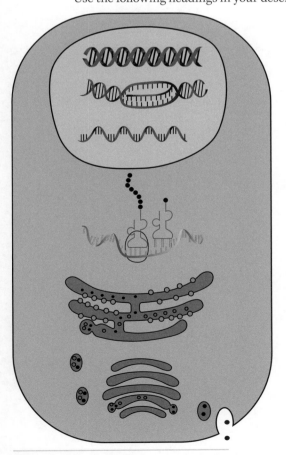

Figure 2.14 The protein secretory pathway

1   Distinguish between proteins that remain in the cell and those that are secreted by the cell. Provide examples of each type.

_____

_____

_____

_____

2   Compare the structure of the endoplasmic reticulum with the structure of the Golgi apparatus.

_____

_____

_____

_____

3   Infer one similarity in structure between the membrane surrounding the Golgi apparatus and the plasma membrane.

_____

_____

_____

2.6 **Chapter review**

2.6.1 Spot the errors

Each diagram in Figure 2.15 contains one or more errors. Circle the errors and write or draw the correct information.

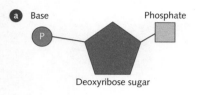

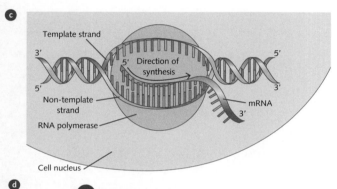

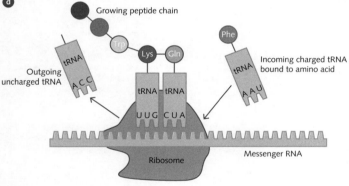

Figure 2.15 Find the errors in these diagrams.

2.6.2 Key terms

Key science skills
Analyse, evaluate and communicate scientific ideas
- use appropriate biological terminology, representations and conventions, including standard abbreviations, graphing conventions and units of measurement

Develop

PAGE 70

Activity: get to know your biology terms

You will need scissors and a glue stick.

Step 1   Cut out the definitions from Figure 2.17 on page 55 by carefully cutting around each shape.

Step 2   Match each definition to its key term in Figure 2.16a & b (pp. 54 and 57), making sure the key term shape fits the definition shape.

Step 3   Glue the definitions in place to make a double helix molecule of key terms and definitions.

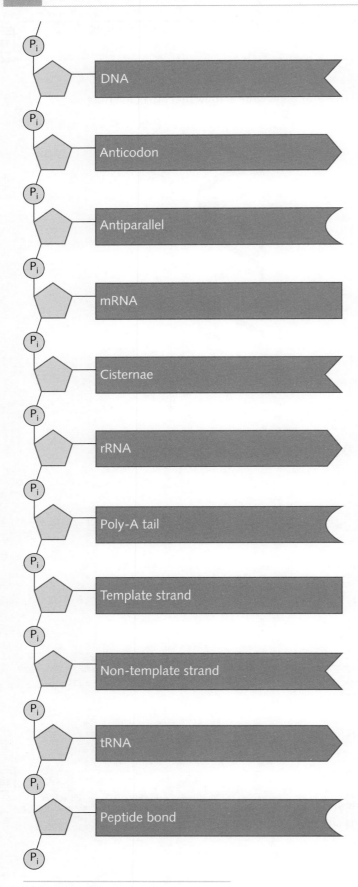

Figure 2.16a

Cut out these definitions and match each one to the correct term in Figure 2.16a and b (pp. 54 and 57) to create your own DNA glossary molecule.

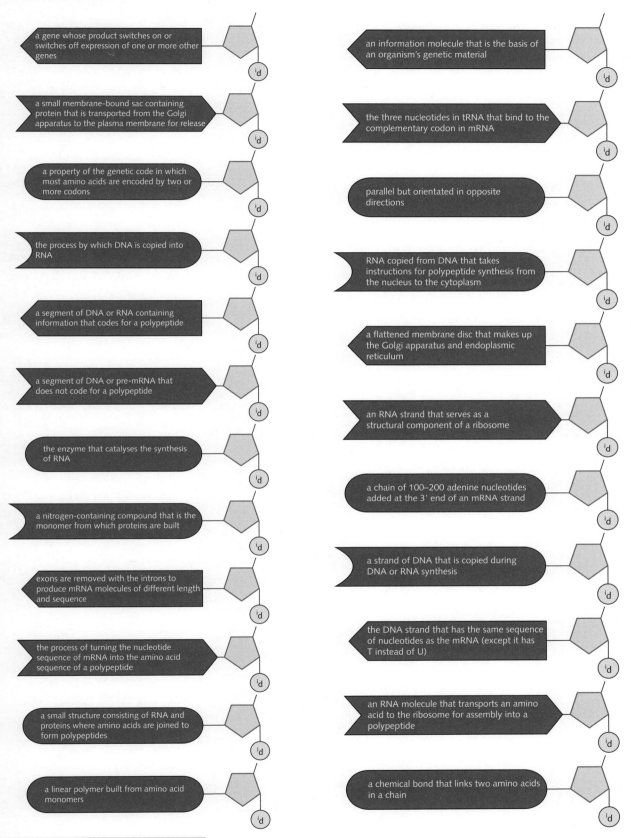

a gene whose product switches on or switches off expression of one or more other genes

a small membrane-bound sac containing protein that is transported from the Golgi apparatus to the plasma membrane for release

a property of the genetic code in which most amino acids are encoded by two or more codons

the process by which DNA is copied into RNA

a segment of DNA or RNA containing information that codes for a polypeptide

a segment of DNA or pre-mRNA that does not code for a polypeptide

the enzyme that catalyses the synthesis of RNA

a nitrogen-containing compound that is the monomer from which proteins are built

exons are removed with the introns to produce mRNA molecules of different length and sequence

the process of turning the nucleotide sequence of mRNA into the amino acid sequence of a polypeptide

a small structure consisting of RNA and proteins where amino acids are joined to form polypeptides

a linear polymer built from amino acid monomers

an information molecule that is the basis of an organism's genetic material

the three nucleotides in tRNA that bind to the complementary codon in mRNA

parallel but orientated in opposite directions

RNA copied from DNA that takes instructions for polypeptide synthesis from the nucleus to the cytoplasm

a flattened membrane disc that makes up the Golgi apparatus and endoplasmic reticulum

an RNA strand that serves as a structural component of a ribosome

a chain of 100–200 adenine nucleotides added at the 3' end of an mRNA strand

a strand of DNA that is copied during DNA or RNA synthesis

the DNA strand that has the same sequence of nucleotides as the mRNA (except it has T instead of U)

an RNA molecule that transports an amino acid to the ribosome for assembly into a polypeptide

a chemical bond that links two amino acids in a chain

Figure 2.17

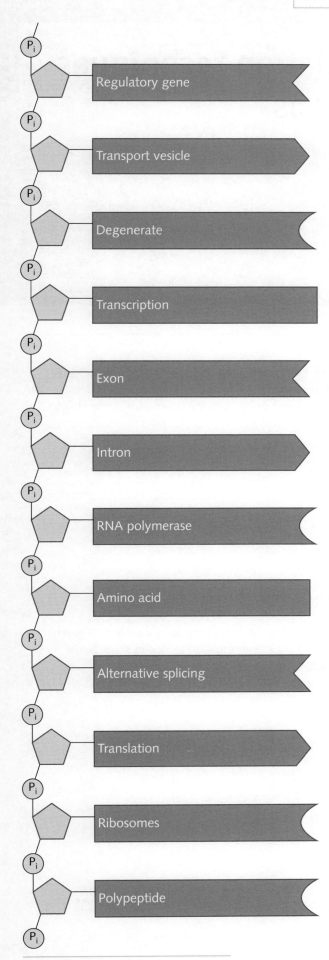

Figure 2.16b

3 DNA manipulation techniques and applications

Remember

In Unit 3 of VCE Biology, you learnt about nucleotides, DNA and how DNA carries the code for proteins. Reflection is key to learning and consolidating new knowledge. Refresh your memory of nucleotides and DNA by answering these questions. This will prepare you to engage deeply with the information in this chapter.

TB
PAGE 80

1 Describe the structure of DNA.

_____

_____

_____

_____

_____

2 Genes are regions of DNA that contain the genetic information for producing a protein. A number of processes are involved in producing a protein from a the information carried in gene. Name these processes in the correct order that they occur within a cell.

_____

_____

_____

3 Explain why enzymes are important in cell functioning.

_____

_____

_____

_____

4 DNA replicates during which two cellular processes?

_____

_____

 9780170452618

3.1   Genetically modified organisms

TB
PAGE 82

Understanding key terms is an important part of learning biology. There are many key terms related to genetically modified organisms that you need to understand and be able to use.

Consolidation of knowledge

Create a visualisation of the key terms below in the space provided. Use a method of your choosing such as a mind map, Venn diagram, flow chart or bubble map. Make sure your visualisation assists you in remembering the key terms and their definitions. Add meaning by augmenting your visualisation with annotations. Add colour to help make these connections more meaningful.

- » Genetic engineering
- » Biotechnology
- » Genetically modified organisms
- » Knock-in organisms
- » Knock-out organisms
- » Transgenic organism
- » Wild type

3.2   Enzymes for modifying DNA

TB
PAGE 84

Key knowledge

DNA manipulation techniques and applications
- the use of enzymes to manipulate DNA, including polymerase to synthesise DNA, ligase to join DNA and endonucleases to cut DNA

One of the essential requirements in genetic engineering is the ability to cut strands of DNA at known sites. The cutting tools used are restriction enzymes. These cut DNA at a specific sequence of nucleotides, known as a restriction site. Restriction enzymes occur naturally in bacteria where they are used to cut the DNA inserted into the bacterial cell from invading viruses.

Consolidation of knowledge

A restriction enzyme binds to its restriction site and cuts the double-stranded DNA at that point. The cut may form either sticky ends, which have overhanging steps that leave some nucleotides exposed (Figure 3.1a), or blunt ends, which have the cut at the same position in each strand (Figure 3.1b).

a T C C A G G A C T G G T A C C G | **A A T T** C C C G G A T A T A T T T C C
A G G T C C T G A C C A T G G C   **T T A A**|G G G C C T A T A T A A A G G

b A C G T C A C T A G|C T T A C C G A
T G C A G T G A T C|G A A T G G C T

Figure 3.1 a A restriction site with sticky ends and **b** a restriction site with blunt ends

Table 3.1 (p. 60) shows some common restriction enzymes (endonucleases) and their recognition sites.

1   In Table 3.1, each enzyme's recognition site is shown using arrows. Fill in the column labelled **After cutting** to show the two cut sections of DNA.

2   In Table 3.1, fill in the column labelled **Blunt or sticky end?** for each enzyme.

Table 3.1 Common restriction endonucleases and their recognition sites

Enzyme	Bacterial source	Recognition site	After cutting	Blunt or sticky end?
Acc16I	*Acinetobacter calcoaceticus 16*	5′ T G C↓G C A 3′ 3′ A C G C G T 5′ ↑		
HacI	*Halococcus acetoinfaciens*	5′ ↓G A T C 3′ 3′ C T A G 5′ ↑		
GstI	*Geobacillus stearothermophilus*	5′ G↓G A T C C 3′ 3′ C C T A G G 5′ ↑		
PacI	*Pseudomonas alcaligenes*	5′ T T A A↓T T A A 3′ 3′ A A T T A A T T 5′ ↑		

3 Which type of end, blunt or sticky, would be best for genetic engineering? Give reasons for your choice.

_____

_____

3.3   # CRISPR-Cas9

PAGE 87

Key knowledge
DNA manipulation techniques and applications
- the function of CRISPR-Cas9 in bacteria and the application of this function in editing an organism's genome

Key science skills
Develop aims and questions, formulate hypotheses and make predictions
- identify, research and construct aims and questions for investigation

Analyse, evaluate and communicate scientific ideas
- analyse and evaluate bioethical issues using relevant approaches to bioethics and ethical concepts, including the influence of social, economic, legal and political factors relevant to the selected issue

Develop

Scientific discoveries fuel even more scientific breakthroughs. For these new discoveries to be meaningful, scientists must be rigorous in process and communication. This task requires you to copy this practice. Read the following information from the US National Library of Medicine.

What are genome editing and CRISPR-Cas9?

Genome editing (also called gene editing) is a group of technologies that give scientists the ability to change an organism's DNA. These technologies allow genetic material to be added, removed or altered at particular locations in the genome. Several approaches to genome editing have been developed. A recent one is known as CRISPR-Cas9, which is short for clustered regularly interspaced short palindromic repeats and CRISPR-associated protein 9. The CRISPR-Cas9 system has generated a lot of excitement in the scientific community because it is faster, cheaper, more accurate, and more efficient than other existing genome editing methods.

CRISPR-Cas9 was adapted from a naturally occurring genome editing system in bacteria. The bacteria capture snippets of DNA from invading viruses and use them to create DNA segments known as CRISPR arrays. The CRISPR arrays allow the bacteria to 'remember' the viruses (or closely related ones). If the viruses attack again, the bacteria produce RNA segments from the CRISPR arrays to target the viruses' DNA. The bacteria then use Cas9 or a similar enzyme to cut the DNA apart, which disables the virus.

The CRISPR-Cas9 system works similarly in the lab. Researchers create a small piece of RNA with a short 'guide' sequence that attaches (binds) to a specific target sequence of DNA in a genome. The RNA also binds to the Cas9 enzyme. As in bacteria, the modified RNA is used to recognise the DNA sequence, and the Cas9 enzyme cuts the DNA at the targeted location. Although Cas9 is the enzyme that is used most often, other enzymes (for example Cpf1) can also be used. Once the DNA is cut, researchers use the cell's own DNA repair machinery to add or delete pieces of genetic material, or to make changes to the DNA by replacing an existing segment with a customised DNA sequence.

Genome editing is of great interest in the prevention and treatment of human diseases. Currently, most research on genome editing is done to understand diseases using cells and animal models. Scientists are still working to determine whether this approach is safe and effective for use in humans. It is being explored in research on a wide variety of diseases, including single-gene disorders such as cystic fibrosis, haemophilia and sickle cell disease. It also holds promise for the treatment and prevention of more complex diseases, such as cancer, heart disease, mental illness, and human immunodeficiency virus (HIV) infection.

Source: U.S. National Library of Medicine

Evidence is one requirement to help support the science behind a new discovery. Using the article above, Figure 3.2 and your own knowledge, outline the steps by answering the questions below that you, as the researcher, could take to show that CRISPR-Cas9 can successfully and safely change the genome of an organism.

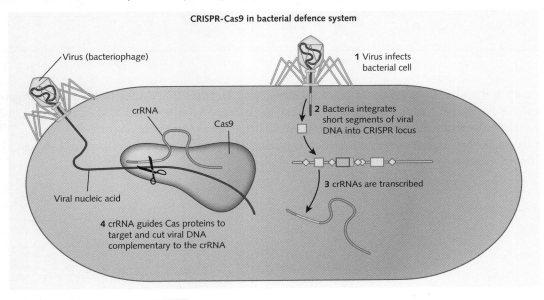

Figure 3.2 CRISPR-Cas9 genome editing

Include the following questions to help you formulate your steps.

1 Aim

What is your research question?

_____

_____

2 Method

a Would you use human subjects in your experiments? What are the alternatives to using humans?

_____

_____

b The researchers who discovered CRISPR-Cas9 did so by accident using a positive control. What is meant by a positive control?

_____

_____

c   Which kind of gene therapy would you use? Would you use germline therapy or somatic therapy? What do each of these mean and what sort of cells do they target? How would the outcomes be different?

d   How would you know if you are successful?

3 Risk assessment

Identify any technical barriers or safety issues when using CRISPR-Cas9. How can these issues be managed?

What are the risks in doing this investigation?	How can these risks be managed to stay safe?

4 Ethics

While some scientists have recognised the potential benefits of using the CRISPR-Cas9 technique to edit genes, other scientists have expressed concern; some are listed below. Use a virtue-based approach and the concepts of beneficence and non-maleficence to answer the following questions. Refer to pages 11 and 12 to revise these terms.

a   Is it acceptable to use gene therapy on an embryo when it is impossible to get permission from the embryo for treatment? Is getting permission from the parents enough?

b   What happens if gene therapies are expensive and only wealthy people can access and afford them?

c   Will some people use genome editing for traits not important for health, such as athletic ability or height? Is that acceptable?

d Edits in germline cells will be passed down through generations. Should scientists ever be able to edit germline cells?

_____

_____

_____

_____

e Identify any social, economic, legal or political factors associated with the concerns you have identified.

Social	
Economic	
Legal	
Political	

3.4 Amplifying DNA

PAGE 92

Key knowledge
DNA manipulation techniques and applications
- amplification of DNA using polymerase chain reaction and the use of gel electrophoresis in sorting DNA fragments, including the interpretation of gel runs for DNA profiling

Key science skills
Analyse, evaluate and communicate scientific ideas
- analyse and explain how models and theories are used to organise and understand observed phenomena and concepts related to biology, identifying limitations of selected models/theories

Develop

Activity: Build a model of the polymerase chain reaction

The polymerase chain reaction (PCR) is a technique used to make many copies of a sample of DNA. It is also called amplification. In this activity you will model the following three steps involved in a polymerase chain reaction.
- Denaturation occurs at 95°C to break the hydrogen bonds between bases.
- Annealing occurs between 50–60°C to enable base pairing and hydrogen bond formation.
- Extension occurs at 72°C, the optimum temperature for the DNA polymerase enzyme.

You will need scissors and a glue stick.

What to do

Figure 3.3 shows a fragment of DNA from a crime scene that requires amplification before it can be analysed. Using this fragment of double-stranded DNA, show what is happening in the three steps of the PCR process. You can either draw or cut out what you need from the material provided in figure 3.4 on page 65.

Annotate the diagrams using labels, arrows and words. Make sure you use correct biological terminology. Use the following words in your explanation.

» Denature
» Temperature
» Primers
» Complementary
» DNA polymerase

» Bases
» Nucleotides
» Hydrogen bonds

Show the product of the PCR process.

Answer the questions below after you have completed the activity.

1 Denaturation

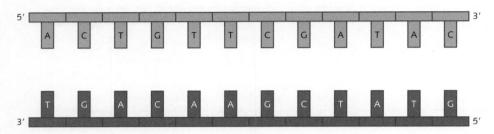

Figure 3.3 Fragment of DNA from a crime scene

2 Annealing

3 Extension

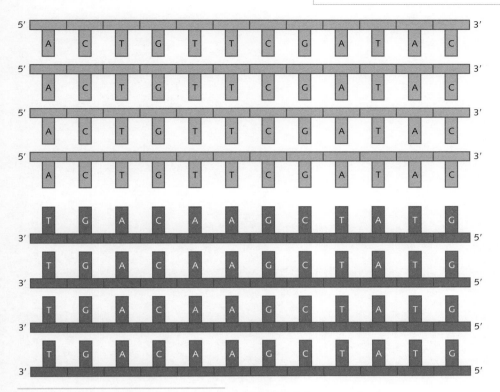

Figure 3.4 Material to be used to build the model of the steps of PCR

After you have built your PCR model, answer the questions below.

1   All scientific models have limitations in what they can show. Describe three features of PCR that are not represented by your model.

_____

_____

_____

2   State one way that you could improve this model.

_____

_____

3   Explain two concepts about PCR that you now understand better as a result of building this model.

_____

_____

_____

3.5   Gel electrophoresis

TB

PAGE 95

Key knowledge

DNA manipulation techniques and applications

• amplification of DNA using polymerase chain reaction and the use of gel electrophoresis in sorting DNA fragments, including the interpretation of gel runs for DNA profiling

Key science skills

Generate, collate and record data

• organise and present data in useful and meaningful ways, including schematic diagrams, flow charts, tables, bar charts and line graphs

Reinforce

Once a sample of DNA has been amplified by PCR, the sample can be sorted by gel electrophoresis. Gel electrophoresis sorts DNA according to the molecular size of the fragment.

DNA from all living organisms has the same chemical components and behaves in the same way during agarose gel electrophoresis. This makes analysis and comparison relatively easy. Scientists can read agarose gels and map the restriction sites of DNA. To prepare a gel electrophoresis run, molten agarose gel is poured with a well comb in place; when the agarose gel is set, the comb is removed to expose the wells so that the DNA fragments can be loaded (Figure 3.5).

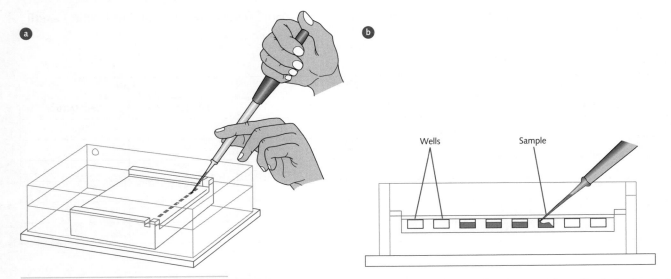

Figures 3.5 a Loading samples of DNA into wells of an agarose gel for gel electrophoresis; **b** side view of samples being loaded

When DNA is being cut for analysis, it is necessary to use a standard DNA fragment for comparison. Bacteriophage lambda (λ)-DNA, a bacterial virus that infects *E. coli*, is often used as a standard piece of DNA. It can be found as either linear or circular molecules. At each end of a linear λ-DNA molecule are single-stranded sequences of 12 nucleotides that are similar to the sticky ends produced by restriction enzymes.

Figure 3.6 shows the results of a standard gel electrophoresis using λ-DNA. Three restriction enzymes, HindIII, EcoRI and BamHI, have been used in a restriction digest (a reaction where restriction enzymes are incubated with DNA in order to cut the DNA at specific sites). The following steps outline the procedure for obtaining the gel shown.

1 Each of the restriction enzymes was placed in a test tube containing a buffer solution and incubated with λ-DNA for 20 minutes.

2 A sample of DNA cut with BamHI was placed into the first well (B) of the agarose gel.

3 A sample of DNA cut with EcoRI was placed into the second well (E) of the agarose gel.

4 A sample of DNA cut with HindIII was placed into the third well (H) of the agarose gel.

5 A sample of DNA mixed with water was placed into the fourth well (–) of the agarose gel.

6 After electrophoresis (an electric current is applied and the negatively charged DNA fragments move towards the positive electrode) and exposure to ultraviolet light (because DNA itself will not be visible in the gel), the banding pattern in Figure 3.6 was obtained. It is possible to determine the approximate size of each fragment of λ-DNA based on how far along the gel it has moved. Smaller fragments will move further along the gel than larger fragments.

What to do

Step 1   Use a ruler to measure the distance from the lower edge of the well to the lower edge of each band on the gel in Figure 3.6. Use your measurements to complete Table 3.2.

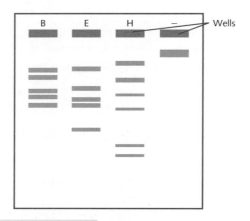

Figure 3.6 In gel electrophoresis, the DNA fragments separate according to molecular size.

Step 2   Enter your measurements in the **Distance** columns. The base pair sizes for fragments cut with HindIII have already been included.

Step 3   Calculate the approximate base pair sizes of the fragments cut with EcoRI and BamHI and complete the **Length** columns.

Table 3.2 Results of gel electrophoresis

HindIII		EcoRI		BamHI	
Distance (mm)	Length (bp)	Distance (mm)	Length (bp)	Distance (mm)	Length (bp)
	23 130		21 226		16 841
	9416				
	6557				
	4361				
	2322				
	2027				

Using your results, answer the following questions.

1 From the bands observed for HindIII, what is the total calculated length of the λ-DNA?

2 The λ-DNA has a known length of 48 502 base pairs. Do the lengths measured from the gel add up to the known length of the λ-DNA? If they do not match, how can you account for the differences in base pair numbers?

3 Why was one of the wells in the gel filled with DNA that had been mixed with water?

4 Why did the band produced by the DNA that was not been cut with a restriction enzyme run the smallest distance through the gel?

5 Describe the process of gel electrophoresis.

6 Explain the role that restriction enzymes play in the process of gel electrophoresis.

7 Summarise the main findings of this investigation.

3.6 DNA profiling

PAGE 98

Key knowledge
DNA manipulation techniques and applications
- amplification of DNA using polymerase chain reaction and the use of gel electrophoresis in sorting DNA fragments, including the interpretation of gel runs for DNA profiling

Key science skills
Generate, collate and record data
- systematically generate and record primary data, and collate secondary data, appropriate to the investigation, including use of databases and reputable online data sources

Analyse and evaluate data and investigation methods
- evaluate investigation methods and possible sources of personal errors/mistakes or bias, and suggest improvements to increase accuracy and precision, and to reduce the likelihood of errors

Develop

Your DNA is very similar to every other human's DNA. Therefore, DNA profiling primarily focuses on the parts of the genetic code that differ most among individuals. These non-coding regions of DNA, called introns, have the most variable coding sequences within members of a species because they do not code for protein.

Forensic scientists use restriction endonucleases to cut these non-coding introns into segments of DNA that contain stretches of repeating nucleotide sequences known as short tandem repeats (STRs). These are unique to each individual and can be used to generate unique DNA fragments that can be investigated. The STRs are separated by gel electrophoresis, and then scientists use the band patterns (created by the fragments) between individuals to determine identity. DNA profiling is used for crime scene investigation, missing person identification, paternity testing, diagnosing genetic disorders, species identification and many other uses. Study each of the two scenarios below and answer the questions below each to show your understanding of DNA profiling.

Scenario 1: Diagnosing genetic disorders

A family of six has a history of a genetic disorder that does not show symptoms until the person is in their 20s. The parents have decided to get all their children tested for this disorder so they can start planning for their futures.

DNA samples were collected from both parents (Figure 3.7). Sample 1 is from the father and sample 2 is from the mother, who is a sufferer of the disease. The children's DNA is in samples 3–6.

1		2		3		4		5		6	
A	T	C	G	A	T	A	T	C	G	A	T
C	G	T	A	A	T	A	T	A	T	A	T
C	G	C	G	G	C	G	C	T	A	G	C
C	G	T	A	C	G	C	G	A	T	C	G
G	C	A	T	T	A	A	T	A	T	T	A
G	C	A	T	T	A	G	C	G	C	T	A
A	T	G	C	C	G	G	C	C	G	G	C
T	A	A	T	C	G	A	T	T	A	C	G
C	G	A	T	A	T	A	T	T	A	C	G
C	G	T	A	T	A	T	A	G	C	C	G
G	C	T	A	G	C	T	A	A	T	C	G
T	A	C	G	T	A	C	G	G	C	G	C
G	C	A	T	G	C	G	C	G	C	G	C
T	A	G	C	G	C	A	T	A	T	G	C
A	T	T	A	A	T	A	T	A	T	A	T
A	T	T	A	T	A	A	T	T	A	T	A
G	C	C	G	C	G	A	T	T	A	C	G
C	G	G	C	C	G	T	A	C	G	C	G
T	A	T	A	G	C	T	A	T	A	T	A
T	A	C	G	A	T	T	A	T	A	G	C
C	G	C	G	G	C	A	T	A	T	G	C

Figure 3.7 DNA samples from the family

1 The geneticist investigating the DNA profiles of these samples used a range of restriction enzymes. Using the sequences of the three restriction enzymes in Table 3.3, highlight where they would be cut in each of the six samples.

Table 3.3 Profiling of DNA sample from a family

Restriction enzyme	Number of cuts	Number of fragments	Length of DNA fragments
BamHI G C G C A T T A C G C G	1. 2. 3. 4. 5. 6.		
HindIII A T A T G C C G T A T A	1. 2. 3. 4. 5. 6.		
EcoRI G C A T A T T A T A C G	1. 2. 3. 4. 5. 6.		

2 Using the data collected, in Table 3.4 draw the banding patterns that would result if these fragments were run on an electrophoresis gel.

Table 3.4 Expected banding patterns from gel electrophoresis

Molecular marker	BamHI	HindIII	EcoRI
21 ●			
20 ●			
15 ●			
10 ●			
5 ●			

3 From the gel run, can you determine if any of the children have inherited the genetic disorder from their mother? Provide evidence from the gel electrophoresis for your answer.

_____

_____

4 What is the advantage of using multiple restriction enzymes to cut the DNA during DNA profiling?

_____

_____

Scenario 2: Paternity testing

A mother seeks to sue a man she once knew for child support expenses. She claims that he is the father of her child. The man vehemently disputes the claim and argues that the child could not be his and he should not have to pay child support. The judge orders for DNA profiling analysis to be done. Study Figure 3.8 showing the DNA profiles of the child, the child's mother and the man and answer the question that follows. (Remember that children receive half of their DNA from their mother and the other half from their father.)

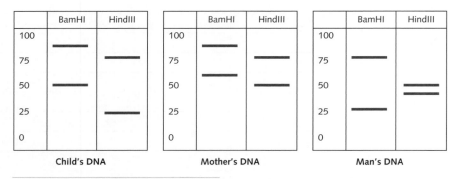

Figure 3.8 DNA profiles of a child, the child's mother and a man

Could the man be the father of this child? Explain your reasoning.

_____

_____

_____

_____

Identifying errors

Although DNA profiling is an effective and powerful tool to tackle cases of missing persons and parentage, there are a number of challenges with DNA profiling in forensic science that are not easy to resolve. Using your knowledge of PCR and gel electrophoresis, evaluate the following ways that DNA profiling could give an unreliable or inaccurate result. Include a discussion of the following points.

a   Personal errors

_____

_____

b   Systematic errors

_____

_____

c   Bias

_____

_____

d   Accuracy

_____

_____

e   Precision

_____

_____

_____

3.7   ## Recombinant plasmids and human insulin

Key knowledge
DNA manipulation techniques and applications
- the use of recombinant plasmids as vectors to transform bacterial cells as demonstrated by the production of human insulin

3.7.1 Recombinant plasmids

Consolidation of knowledge   PAGE 106

1   Gene cloning uses bacterial plasmids to produce many copies of a gene. The overall process is summarised in the flow chart in Figure 3.9. At each step in the flow chart, add more detail and diagrams in the space provided. Remember to name the enzymes involved.

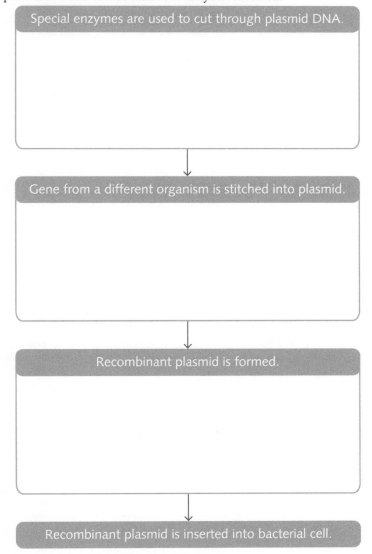

Special enzymes are used to cut through plasmid DNA.

↓

Gene from a different organism is stitched into plasmid.

↓

Recombinant plasmid is formed.

↓

Recombinant plasmid is inserted into bacterial cell.

Figure 3.9 Gene cloning

2   Explain why bacterial plasmids are used in gene cloning.

_____

_____

_____

3   What features of DNA allow it to be translated when in a bacterial cell?

_____

_____

_____

_____

3.7.2 Synthesising the human insulin gene

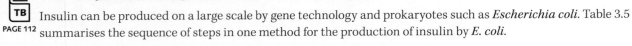

Insulin can be produced on a large scale by gene technology and prokaryotes such as *Escherichia coli*. Table 3.5 summarises the sequence of steps in one method for the production of insulin by *E. coli*.

Consolidation of knowledge

1   Complete Table 3.5 by adding one statement to each of the empty boxes.

Table 3.5 The sequence of steps in one method for the production of insulin by _E. coli_

Step		Reason for step
1	Obtain copies of genes with sticky ends	The gene codes for the synthesis of insulin
2		Acts as a vector for the transfer of the gene into the host
3	Use the same endonuclease enzyme on the plasmid as was used on insulin	
4	Mix vector and gene	
5		To seal the sugar–phosphate backbone
6		To obtain transformed host *E.coli* cells
7	Screen for, and obtain, successfully transformed cells	
8		To obtain large amounts of insulin for extraction and purification

2 Explain two advantages of treating diabetes with human insulin produced by gene technology rather than using insulin from animals.

_____

_____

_____

_____

3 Suggest why it would be beneficial to genetically modify a:

a structural gene.

_____

_____

b regulatory gene.

_____

_____

3.7.3 Plasmid X

Key science skills

Analyse, evaluate and communicate scientific ideas

- analyse and explain how models and theories are used to organise and understand observed phenomena and concepts related to biology, identifying limitations of selected models/theories

TB
PAGE 112
Develop

An artificial plasmid known as plasmid X has been constructed to act as a vector and has been used to insert the human insulin gene into the bacterium *Escherichia coli*. Plasmid X included two antibiotic-resistance genes (an ampicillin-resistance gene and a tetracycline-resistance gene) and a target site for the BamHI restriction enzyme in the middle of the tetracycline-resistance gene.

Plasmid X was cut with BamHI and the gene for human insulin was inserted into it (Figure 3.10).

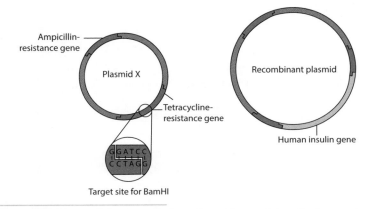

Figure 3.10 Plasmid X and the recombinant plasmid containing the human insulin gene

The bacteria were then mixed with the recombinant plasmids. The bacteria that had successfully taken up recombinant plasmids were identified using the following steps.

Step 1 Bacteria were spread onto Petri dishes containing nutrient agar and ampicillin and were incubated to allow colonies to form.

Step 2 Some of the bacteria from each of the colonies were transferred to other Petri dishes containing nutrient agar and tetracycline, as shown in Figure 3.11.

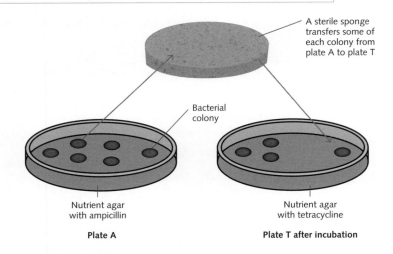

Figure 3.11 The process for transferring bacteria from one Petri dish (A) onto another (T)

1   Using your knowledge of antibiotic selection, explain why the bacteria were first spread onto plates containing ampicillin.

_____

_____

_____

2   Explain why it is important that the target site for BamHI on plasmid X is in the middle of the tetracycline-resistance gene to be able to identify bacteria that have successfully taken up the recombinant plasmid.

_____

_____

_____

3   On Figure 3.11 use a label line and the letter R to identify a colony of bacteria that contain the recombinant plasmid.

4   Plasmid vectors carrying antibiotic resistance genes are now rarely used in gene technology. Explain why antibiotic-resistance genes are now rarely used.

_____

_____

5   Suggest one type of gene that has replaced antibiotic-resistance genes in plasmid vectors. (Hint: Think of glow in the dark.)

_____

_____

3.8 Genetic engineering in agriculture

Key knowledge
DNA manipulation techniques and applications
- the use of genetically modified and transgenic organisms in agriculture to increase crop productivity and to provide resistance to disease

3.8.1 Literature review

TB
PAGE 115

Develop

Key science skills
Construct evidence-based arguments and draw conclusions
- distinguish between opinion, anecdote and evidence, and scientific and non-scientific ideas

The literature review process is an important aspect of scientific discovery. In order to assert that something is new, innovative or novel or to make claims about an existing science, you must know the field. A literature review establishes the legitimacy of any claims you make by giving readers a clear view of where your position fits within the science you are talking about and how valid your claims are.

A Year 12 Biology class was asked to undertake a literature review of the following bioethical issue and to form their own opinion on that issue.

To meet the growing needs of food for the increasing population, some countries use genetically modified foods. However, some people believe that genetically modified foods are not only unhealthy but affect nature too. Do you agree or disagree with this?

Benjamin wrote the following report. Read his report and evaluate its scientific rigour. Then answer the questions that follow, which are designed to help you form an assessment of this issue and communicate it to others.

My Science Report: GM foods are bad!

With the rapid increase in population resulting in the never-ending demand for food, some section of people have the opinion that genetically modified (GM) food is the answer, while others believe that it might actually be a curse for the masses and the environment. I assert that altering the gene make-up of plants, using biotechnology, might prove more harmful than a boon to human race. This report will discuss this issue, using relevant examples to demonstrate points and support arguments.

Firstly, from the economic perspective, GM food could be costly. Furthermore, pesticides and herbicides use increases. These pesticides are poisonous for the people as well as for the wildlife. For instance, in South America, pesticide spraying is causing serious health problems for the people as well as for other forms of life. Consequently, this might culminate into some chronic diseases or it might even lead to new diseases in the future generation.

Secondly, long-term effects of GM foods are still a matter of research. In addition, they might prove to be toxic to certain organisms, such as bees and butterflies. For example, research has shown that bees, being an important pollinator of many food crops, might endanger their existence. Therefore, many developing countries across the world are moving towards ecological farming, which might be a better and safer way to feed the poor and hungry.

To conclude, I believe that GM food does affect our health and wellbeing in spite of Food Standards Australia claiming that they are safe. The claim of ending the world hunger is false, as it is not caused by shortage of food production but by sheer mismanagement and lack of access to food brought about by various social and political causes.

By *Benjamin*

1 Is Benjamin's answer supported by scientific data, biased or an opinion piece? Give reasons for your answer.

2   Select an appropriate research methodology to gather evidence to support or reject the following statements by Benjamin.

   a   'GM food could be costly'

_____

   b   'Pesticides and herbicides use increases'

_____

   c   'Long-term effects of GM foods are still a matter of research'

_____

   d   'In spite of Food Standards Australia claiming that they are safe'

_____

3   Benjamin's report is incomplete. List the essential elements of a scientific report and state what is missing from Benjamin's report.

_____

_____

_____

_____

_____

_____

PAGE 123

3.9 Chapter review

3.9.1 Key terms

To complete this task you will need scissors and a glue stick.

Figure 3.12 shows some definitions of key terms that you have met during your study of DNA manipulation techniques and applications. Cut out the definitions and match them to their key terms in Figure 3.13 on page 81 to make either blunt or sticky ends.

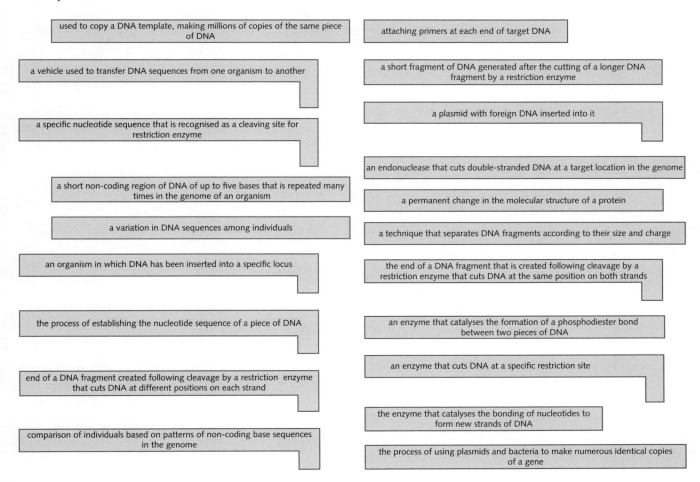

used to copy a DNA template, making millions of copies of the same piece of DNA

attaching primers at each end of target DNA

a vehicle used to transfer DNA sequences from one organism to another

a short fragment of DNA generated after the cutting of a longer DNA fragment by a restriction enzyme

a specific nucleotide sequence that is recognised as a cleaving site for restriction enzyme

a plasmid with foreign DNA inserted into it

a short non-coding region of DNA of up to five bases that is repeated many times in the genome of an organism

an endonuclease that cuts double-stranded DNA at a target location in the genome

a variation in DNA sequences among individuals

a permanent change in the molecular structure of a protein

an organism in which DNA has been inserted into a specific locus

a technique that separates DNA fragments according to their size and charge

the process of establishing the nucleotide sequence of a piece of DNA

the end of a DNA fragment that is created following cleavage by a restriction enzyme that cuts DNA at the same position on both strands

an enzyme that catalyses the formation of a phosphodiester bond between two pieces of DNA

end of a DNA fragment created following cleavage by a restriction enzyme that cuts DNA at different positions on each strand

an enzyme that cuts DNA at a specific restriction site

comparison of individuals based on patterns of non-coding base sequences in the genome

the enzyme that catalyses the bonding of nucleotides to form new strands of DNA

the process of using plasmids and bacteria to make numerous identical copies of a gene

Figure 3.12 Cut out these definitions and match them to the key terms on page 81. When you have matched them, glue them in place to make either blunt or sticky ends.

Figure 3.13 Match these key terms to the corresponding definitions in Figure 3.12 to make either blunt or sticky ends

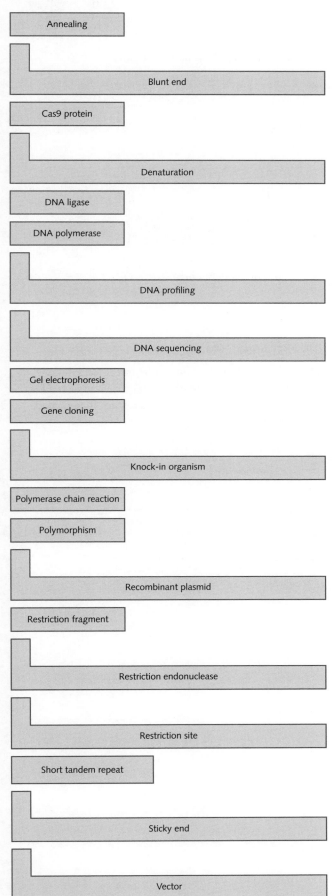

Chapter review (continued)

3.9.2 Exam practice

PAGE 125

Exam practice

1 ©VCAA 2019 Section A Q37 (adapted) MEDIUM Bt corn expresses a protein that acts as an insecticide.

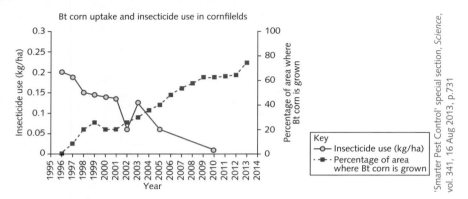

'Smarter Pest Control' special section, *Science*, vol. 341, 16 Aug 2013, p.731

Based on your knowledge and the data in the graph, what is a benefit of using Bt corn?

A The area planted with Bt corn is decreasing.

B Less insecticide is used with Bt corn crops.

C Bt corn is cheaper to produce than non-Bt corn.

D Bt corn is more expensive to produce than non-Bt corn.

2 ©VCAA 2018 Section A Q34 (adapted) EASY Genetic testing can be used to test for the allele for Huntington's disease. The onset of Huntington's disease predominantly occurs in adulthood. Eight members of a family were tested for the Huntington's disease allele. The diagram below shows the electrophoresis gel results of a test for the presence of the allele. Individuals 4 and 8 have been diagnosed with the disease.

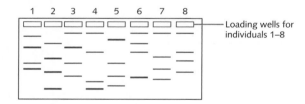

Which other individual is likely to suffer from Huntington's disease, now or in the future?

A 1

B 3

C 6

D 7

3 ©VCAA 2017 Section A Q39 (adapted) MEDIUM DNA profiling using short tandem repeats (STRs) helps to determine the genetic relationship between individuals. DNA profiles based on four STRs for five individuals are shown below. The results of a gender identifier are also shown.

STR	Individual 1	Individual 2	Individual 3	Individual 4	Individual 5
CSF1PO	7,14	7,11	8,13	7,14	7,14
TPOX	6,10	10,12	6,9	10,12	10,10
D21S11	27,30	29,32	27,27	29,30	27,28
D8S1179	9,11	12,13	17,17	11,12	9,11

	Individual 1	Individual 2	Individual 3	Individual 4	Individual 5
Gender identifier	Male	Female	Female	Male	Male

Which one of the following conclusions can be made using the information given?

A Individual 2 is the mother of individual 3.

B Individuals 1 and 2 could be the parents of individual 4.

C Individual 4 is the father of individual 5.

D Individuals 1 and 3 could be the parents of individual 5.

4 ©VCAA 2019 Section B Q8 (adapted) The Genomics Health Futures Mission will run a $32 million trial, starting in 2024, to screen more than 10 000 couples who are in early pregnancy or who are planning to have a baby. A blood test will be used to screen for 500 severe or fatal gene mutations. Couples will be told they have a genetic mutation if both individuals in the couple carry the same mutation. The trial may lead to a population-wide carrier screening program. The researchers will evaluate cost effectiveness, psychological impact, ethics and barriers to screening. It is anticipated that future tests will be free of charge.

a Explain why couples will be told they have a genetic mutation only if both individuals carry the same mutation. ©VCAA EASY

1 mark

b The blood sample from an individual will provide researchers with only a small amount of DNA. Name the technique that will be used to increase the amount of DNA available to researchers. State the names of the stages in this technique, and the temperature at which each stage must be performed and why. ©VCAA HARD

4 marks

c Once the DNA has been amplified, it can be loaded into a well of an agarose gel. State the name of this process and explain how it works to achieve a result. ©VCAA HARD

3 marks

d   A couple find out that they both carry a severe genetic mutation of the same gene but have an unaffected baby girl. What is one ethical and one social issue that faces the parents when that baby girl becomes of child-bearing age? ©VCAA HARD   2 marks

	Issue or implication
Ethical	
Social	

Enzymes and the regulation of biochemical pathways

4

Remember

TB
PAGE 136

In Unit 1 of VCE Biology, you learnt about cell structure. In Unit 3 of VCE Biology, you learnt about protein synthesis, including the synthesis of enzymes. This knowledge will help you work though this chapter of the workbook. Test yourself by answering these questions from memory or complete this section to solidify your knowledge as you work through the chapter.

1 What is the role of enzymes in a biochemical pathway?

_____

_____

2 What is the role of photosynthesis?

_____

_____

3 In which organelle of a plant cell does photosynthesis occur?

_____

4 What is the role of cellular respiration?

_____

_____

5 In which organelle of a eukaryotic cell does cellular respiration occur?

_____

4.1 Biochemical pathways for cell metabolism

Key knowledge
Regulation of biochemical pathways in photosynthesis and cellular respiration
- the general structure of the biochemical pathways in photosynthesis and cellular respiration from initial reactant to final product
- the general role of enzymes and coenzymes in facilitating steps in photosynthesis and cellular respiration

Consolidation
of knowledge

4.1.1 Organising key terms

PAGE 137 Using the correct terms is critical for effective scientific communication. You will find it useful to spend time now sorting, clarifying and categorising the beaker of terms shown in Figure 4.1, when it comes to understanding biological chemical reactions.

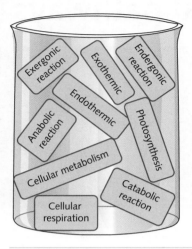

Figure 4.1 Key terms

1 Organise the key terms in Figure 4.1 by writing them into Figure 4.2 to construct a diagrammatic understanding of their meanings.

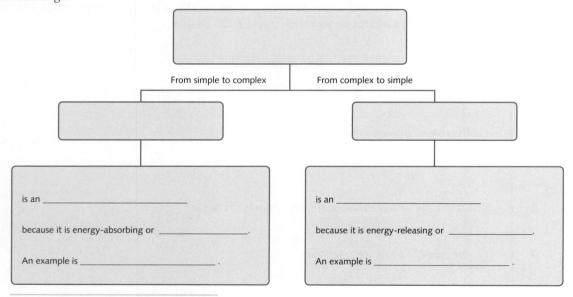

Figure 4.2 Key terms and their meanings

2 Figure 4.3 shows the relationship between enzymes, cofactors and their substrates. Show your understanding of this relationship by labelling each figure with the relevant terms below.

- Reactant/substrate
- Enzyme–substrate complex
- Product(s)
- Enzyme
- Active site
- Cofactor
- Substrate

Figure 4.3 Enzymes, cofactors and their substrates

  9780170452618

3 Biochemical pathways involve a series of enzymes and coenzymes or cofactors, which are molecules that help an enzyme function properly. Colour is a great way to assist the recall of information. Colour in the illustration in Figure 4.4 to help you memorise this structure and understand its function.

Use the following colours to illustrate the different parts of the structure:

Substrate – green
Enzyme – blue
Cofactor – orange
Product – yellow

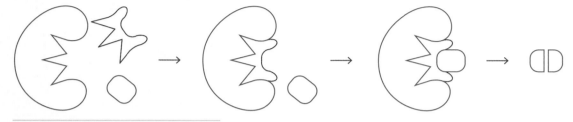

Figure 4.4 Biochemical pathways

4 Now it is time to consolidate your knowledge about coenzymes and cofactors. Describe the function of a coenzyme and a cofactor in relation to the correct functioning of an enzyme.

4.1.2 Biochemical pathways and coenzymes

Key science skills
Develop aims and questions, formulate hypotheses and make predictions
- identify, research and construct aims and questions for investigation
- formulate hypotheses to focus investigation
- predict possible outcomes

Generate, collate and record data
- organise and present data in useful and meaningful ways, including schematic diagrams, flow charts, tables, bar charts and line graphs.

Reinforce
TB
PAGE 138

This activity explores the reactions in a biochemical pathway and the function of coenzymes. Glycolysis is a biochemical pathway that has several intermediate reactions to produce pyruvate. The first step in this pathway is shown in Figure 4.5.

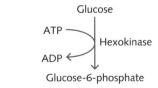

Figure 4.5 The first step in the glycolysis pathway

1 ATP is a coenzyme. How would the absence of this coenzyme affect the end product of this biochemical pathway?

2 Is the part of the pathway shown in Figure 4.5 endothermic or exothermic? How can you tell?

3 How does ATP provide the energy to provide energy for a chemical reaction to proceed? Use the space below to draw a diagram to show your understanding.

4 Figure 4.6 shows a hypothetical biochemical pathway.

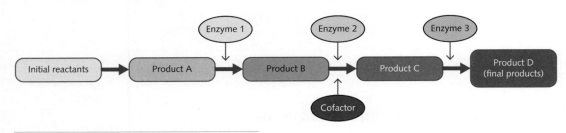

Figure 4.6 Hypothetical biochemical pathway

a What would be the product of this biochemical pathway if enzyme 1 was not available?

_____

_____

b What would be the product of this biochemical pathway if the cofactor were not available?

_____

_____

c If enzymes 1 and 3 and the cofactor were available, but enzyme 2 was not available, what would be the product of this biochemical pathway?

_____

4.1.3 Scientific literacy

Key science skills

Analyse, evaluate and communicate scientific ideas

- critically evaluate and interpret a range of scientific and media texts (including journal articles, mass media communications and opinions in the public domain), processes, claims and conclusions related to biology by considering the quality of available evidence

Develop

In order to present credible ideas, scientists must first understand where their research is situated within their field of inquiry. A critical skill is the ability to locate and validate credible research. You will now engage in this process. Use your critical eye to determine the value, relevance and credibility of the following article.

Scientists construct energy production unit for a synthetic cell

'Our aim is the bottom-up construction of a synthetic cell that can sustain itself and that can grow and divide,' explains University of Groningen Professor of Biochemistry Bert Poolman. He is part of a Dutch consortium that obtained a Gravitation grant in 2017 from the Netherlands Organisation for Scientific Research to realise this ambition. Different groups of scientists are producing different modules for the cell and Poolman's group was tasked with energy production.

All living cells produce ATP as an energy carrier but achieving sustainable production of ATP in a test tube is not a small task. 'In known synthetic systems, all components for the reaction were included inside a vesicle. However, after about half an hour, the reaction reached equilibrium and ATP production declined,' Poolman explains. 'We wanted our system to stay away from equilibrium, just like in living systems.'

It took three PhD students in his group nearly four years to construct such a system. A lipid vesicle was fitted out with a transport protein that could import the substrate arginine and export the product ornithine. Inside the vesicle, enzymes were present that broke down the arginine into ornithine. The free energy that this reaction provided was used to link phosphate to ADP, forming ATP. Ammonium and carbon dioxide were produced as waste products that diffused through the membrane. 'The export of ornithine produced inside the vesicle drives the import of arginine, which keeps the system running for as long as the vesicles are provided with arginine,' explains Poolman.

Source: University of Groningen. 'Scientists construct energy production unit for a synthetic cell.' *ScienceDaily*, 18 September 2019

Having closely read this article, answer the following four questions, including determining the article's credibility.

1 State the aim of this research.

_____

_____

2 Use the information in the extract to construct a word equation to show the substrate, enzymes and products in this biochemical pathway. Include any other relevant information.

3 State two factors that could affect the forward progress of this reaction.

_____

_____

4 Provide two reasons why you think this article comes from a reputable source.

_____

_____

_____

_____

4.2   Enzymes: the key to controlling biochemical pathways

Key knowledge
Regulation of biochemical pathways in photosynthesis and cellular respiration
*   the general structure of the biochemical pathways in photosynthesis and cellular respiration from initial reactant to final product
*   the general role of enzymes and coenzymes in facilitating steps in photosynthesis and cellular respiration

4.2.1 Specificity of enzymes

PAGE 142

Key science skills
Analyse, evaluate and communicate scientific ideas
*   analyse and explain how models and theories are used to organise and understand observed phenomena and concepts related to biology, identifying limitations of selected models/theories

Develop

The following paragraph gives an insight into how scientific theories develop over time. Read the paragraph and answer the questions that follow about the two theories of enzyme action.

Enzymes are proteins that catalyse biochemical processes. In 1894, Emil Fischer proposed a theory known as the 'lock and key model' to explain the actions of enzymes and the way in which they fit with substrates. In 1959, Daniel E Koshland Jr proposed another theory known as the 'induced fit model' in which he explained that the active site was not static but changed when the enzyme attaches.

Answer the following five questions to investigate how two competing theories could have existed at the same time.

1   Describe the lock and key model of enzyme action.

_____

_____

_____

2   Describe the induced fit model of enzyme action.

_____

_____

_____

3   Contrast these two models. (How are they different?)

_____

_____

_____

_____

4   Theories are constantly being reviewed and changed as more evidence becomes available. Explain why the lock and key model is no longer thought to be the best model of enzyme function.

_____

_____

_____

_____

_____

5   Use an analogy to explain the induced fit model of enzyme function.

_____

_____

_____

_____

_____

_____

4.2.2 Enzymes need help: coenzymes and cofactors

1   Write a definition of the following terms relating to coenzymes and cofactors. Make sure you state whether the coenzymes are loaded or unloaded and which biochemical reactions they participate in.

Consolidation of knowledge   **TB PAGE 143**

a   Loaded

_____

_____

b   Unloaded

_____

_____

c   Coenzyme

_____

_____

d   Substrate

_____

_____

e   ATP

_____

_____

f   ADP

_____

_____

g $NADP^+$

_____

_____

h   NADPH

_____

_____

i   NAD⁺

j   NADH

k   FAD

l   FADH$_2$

2   Identify a clue to help you remember whether a coenzyme is loaded or unloaded.

3   Contrast coenzymes with cofactors.

Key knowledge
Regulation of biochemical pathways in photosynthesis and cellular respiration
- the general factors that impact on enzyme function in relation to photosynthesis and cellular respiration: changes in temperature, pH, concentration, competitive and non-competitive enzyme inhibitors

4.3.1 Effect of temperature

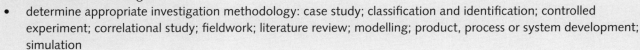

Key science skills
Develop aims and questions, formulate hypotheses and make predictions
- identify independent, dependent and controlled variables in controlled experiments
- formulate hypotheses to focus investigation
- predict possible outcomes

Plan and conduct investigations
- determine appropriate investigation methodology: case study; classification and identification; controlled experiment; correlational study; fieldwork; literature review; modelling; product, process or system development; simulation

Enzymes help to speed up biochemical reactions without being used up in the process. They are found in all living cells. Enzymes are proteins, and therefore a range of factors can contribute to how effectively they operate in biochemical reactions.

A practical investigation was conducted to look at the effect of temperature on the activity of the enzyme diastase, which is found in most living cells. Diastase acts on starch to convert it into maltose. The investigators looked for the presence or absence of starch to measure the activity of diastase. They did this by adding iodine to the solution; if starch is present, the solution will turn a blue-black colour.

Use the information above to help you answer the following questions.

1   Formulate a research question to investigate the effect of temperature on the activity of an enzyme.

2   Formulate a hypothesis to investigate the effect of temperature on the activity of an enzyme.

3   State the:

   a   independent variable.

   b   dependent variable.

4   What methodology would you choose to test your research question? Explain your choice.

5   The investigation looked at five different temperatures (0, 10, 30, 45 and 60°C) and three different volumes of diastase solution (1, 2 and 5 mL). Is this an appropriate method? Rewrite the method to ensure that it will test the hypothesis and lead to accurate results so that valid conclusions can be drawn.

6   Consider what you know about the activity of enzymes in cells. Draw a graph predicting the rate of enzyme activity against temperature.

7   Given that enzymes are proteins, what is the likely effect of high temperature on their structures? How will this affect their functions?

_____

_____

_____

_____

4.3.2 Effect of changing pH

PAGE 145

Two Biology students were investigating the effect of pH on the activity of three different enzymes in humans. Their results are shown in Table 4.1. Spend some time now to make the connection between the aim of this investigation and the data collected below. Having a clear understanding of this relationship will help you answer the questions that follow.

Table 4.1 Effect of pH on enzyme activity

pH	Enzyme activity (%)		
	Amylase	Pepsin	Arginase
1	0	50	0
2	0	100	0
3	0	50	0
4	0	20	0
5	35	5	0
6	65	0	20
7	100	0	40
8	65	0	60
9	35	0	80
10	0	0	100
11	0	0	80
12	0	0	60

1   Formulate the possible research question that the students were investigating.

_____

_____

2   Formulate the possible hypothesis that the students were investigating.

_____

_____

3   State the:

   a   independent variable. _____

   b   dependent variable. _____

4   Use the following grid to graph the data. Remember to include all the features of a good graph.

5 Do the results support or refute the hypothesis under investigation? Provide evidence to support your answer.

_____

_____

_____

_____

6 State the optimum pH for each of the following enzymes.

a Amylase _____

b Pepsin _____

c Arginase _____

7 Considering the optimum pH for each enzyme, where would you reasonably expect to find each of the following enzymes in the human body?

a Amylase _____

b Pepsin _____

c Arginase _____

8 Write a conclusion of no more than 60 words. In your conclusion you need to:

- restate the hypothesis
- state whether the hypothesis is supported or refuted
- state the main findings
- recommend what to investigate next as a result of your findings.

_____

_____

_____

_____

_____

_____

_____

4.3.3 Effect of substrate and enzyme concentration

Key science skills
Develop aims and questions, formulate hypotheses and make predictions
*   identify, research and construct aims and questions for investigation
*   identify independent, dependent and controlled variables in controlled experiments
*   formulate hypotheses to focus investigation
*   predict possible outcomes

Practise

PAGE 146

An investigation was conducted in which proteins were mixed with the enzyme protease. Proteases catalyse the breakdown of proteins. A group of students wanted to test which of the following caused the protein to break down the fastest: the amount of protein or the amount of enzyme. Use this information to answer the following questions.

1   If the enzyme successfully catalysed the breakdown of proteins, what substances would the students be testing for in the test tubes?

_____

2   Formulate the possible research question that the students were investigating.

_____

_____

3   Formulate the possible hypothesis that the students were investigating.

_____

_____

4   State the:
    a   independent variable.   _____
    b   dependent variable.   _____
5   State any controlled variables.

_____

_____

_____

6   Describe a control for this investigation.

_____

_____

7   Predict which factor (amount of protein or amount of enzyme) you think will have the biggest impact on the rate of reaction. What results would support this prediction?

_____

_____

_____

_____

_____

8   The teacher suggested that the size of the piece of protein could affect the reaction rate. Design a method to test this suggestion.

_____

_____

_____

_____

_____

_____

_____

_____

_____

4.3.4 Enzyme inhibitors

Non-competitive inhibitors are molecules that bind to a part of an enzyme that is not the active site. This is called the allosteric site. The non-competitive inhibitor acts to change the formation of the enzyme and the active site. Competitive inhibitors compete directly with the substrate for space in the active site and prevent the substrate from binding. Competitive inhibitors either lower the reaction rate or stop the reaction from taking place.

Consolidation of knowledge

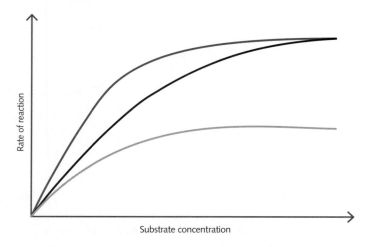

Figure 4.7 Rate of reaction by substrate concentration

1   Complete the graph in Figure 4.7 by using the following labels to label the lines on the graph and demonstrate your understanding of these processes.

- Normal enzyme reaction
- Reaction with non-competitive inhibitor
- Reaction with competitive inhibitor

2 Justify your labelling, with reference to the action of inhibitors.

_____

_____

_____

_____

_____

_____

_____

_____

_____

4.4 Chapter review

4.4.1 Key terms

PAGE 154 The table below shows the definitions of a few key terms relating to enzymes. Write the key terms that go with the corresponding definitions on the column on the right.

Definition	Key term
A short-term molecule carrier of energy within the cell	
An unloaded coenzyme	
A loaded coenzyme	
Secondary binding site on the enzyme that a non-competitive inhibitor binds to	
Molecule with a similar chemical structure to the enzyme's substrate	
Speed at which a biochemical reaction occurs	
One model of enzyme function	
Place on the surface of an enzyme molecule where substrate molecules attach	
Atoms are joined to make more complex molecules	
Variable that limits the rate of a reaction	
Sum of metabolic reactions in a cell	
Energy required to initiate a reaction	
Enzyme that provides energy for the cell through synthesis of ATP	
Anabolic reaction using light energy to form glucose from water and carbon dioxide	
Series of chemical reactions, each controlled by an enzyme, that converts a substrate molecule to a final product	

Figure 4.8 Table of key terms and definitions. Use your completed table for revision later.

Chapter review continued

4.4.2 Exam practice

TB
PAGE 155

Exam
practice

The following graph illustrates the effect of different concentrations of a substrate on the rate of a cellular reaction.

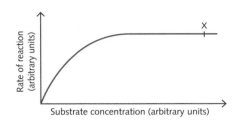

1 ©VCAA 2003 E1 Section 1 Q17 (adapted) HARD In this series of experiments, the amount of enzyme, the pH and the temperature remained constant.

At point X, it would be reasonable to conclude that the

A solution has become too acidic, which is stopping the reaction proceeding.
B heat released from the reaction has denatured the enzyme.
C active sites on the enzyme are all full.
D reaction has run out of substrate.

2 ©VCAA 2004 E1 Section 1 Q7 (adapted) HARD Scientists have found micro-organisms called extremophiles living in springs of highly acidic water. Many of these micro-organisms have been classified as prokaryotic. The enzyme activity of these micro-organisms was investigated over a range of pH values. The results are shown in the following graph.

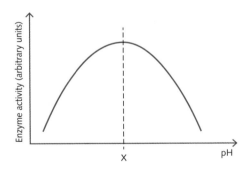

The pH at X is most likely to be
A 3.
B 5.
C 7.
D 10.

3 ©VCAA 2002 E1 Section 1 Q25 (adapted) EASY The rate of reaction of a typical human enzyme was compared with the rate of reaction of a typical enzyme taken from bacteria that live in hot springs. The rates of reaction were measured over the same range of temperatures. The results are shown in the following graph.

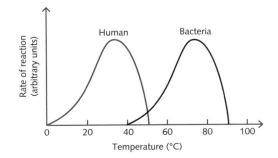

It is reasonable to conclude that

A   both human and bacterial enzymes can operate at 45°C.

B   typical bacterial enzymes fail to act at temperatures below 50°C.

C   a denatured human enzyme would resume activity if incubated at 20°C.

D   enzymes from bacteria that live in hot springs withstand temperatures up to 100°C.

4   ©VCAA   2002 E1 Section 1 Q24 (adapted) MEDIUM   Nine tubes containing the same amount of enzyme and the same volume of a solution with different concentrations of substrate (glucose) were incubated under identical conditions. The rate of break down of glucose in each tube was measured and plotted against the original concentration of the substrate. The results are shown in the following graph.

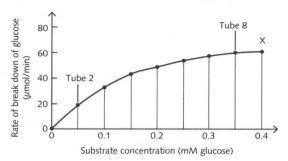

It is reasonable to conclude that

A   no control tube exists in this experiment.

B   the rate of breakdown of glucose in tube 8 is 40 μmol/min.

C   the low rate of reaction in tube 2 is due to a low substrate concentration in that tube.

D   the graph tapers off at point X because all the enzyme has been used in tube 6.

5   ©VCAA   2006 E1 Section 1 Q25 (adapted) MEDIUM   In the production of isoleucine from threonine in bacteria (Biochemical pathway 1), the end product acts as an inhibitor of the first enzyme in that pathway. In the production of arginine (Biochemical pathway 2), the end product has no influence on other enzymes in that pathway.

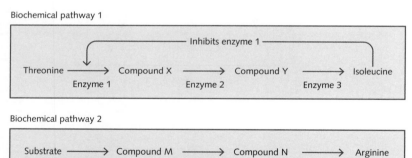

It is reasonable to conclude that in

A   Biochemical pathway 1, if the production of enzyme 2 stops there would be continuous production of isoleucine.

B   Biochemical pathway 2, if the production of enzyme 1 stops there would be continuous production of arginine.

C   Biochemical pathway 1, providing all enzymes are present, the break down of threonine would be continuous if there was a continuous supply of isoleucine.

D   Biochemical pathway 2, providing all enzymes are present, the production of arginine would be continuous if there was a continuous supply of substrate.

Biochemical pathways: photosynthesis and cellular respiration

5

Remember

TB
PAGE 162

In Unit 1 of VCE Biology, you learnt about chloroplasts and mitochondria. In Unit 3 of VCE Biology, you learnt about biochemical pathways and enzyme functions. There are links that you will need to make between the knowledge indicated above and the new knowledge you will be introduced to in this section. Answer the questions below as a way to revise what you know already. Use these questions to reflect on how one set of ideas might relate to the new ideas you are about to encounter.

Photosynthesis

1   Where does the energy required for photosynthesis come from?

2   Where in the cell does photosynthesis take place?

3   What are the inputs and outputs of photosynthesis?

Cellular respiration

4   Where in the cell does cellular respiration take place?

5   What are the inputs and outputs of cellular respiration?

6   What is the name of the energy-storing molecule that is produced during cellular respiration?

5.1   Photosynthesis as a biochemical pathway

Key knowledge
Photosynthesis as an example of biochemical pathways
- inputs, outputs and locations of the light dependent and light independent stages of photosynthesis in C_3 plants (details of biochemical pathway mechanisms are not required)
- the role of Rubisco in photosynthesis, including adaptations of C_3, C_4 and CAM plants to maximise the efficiency of photosynthesis
- the factors that affect the rate of photosynthesis: light availability, water availability, temperature and carbon dioxide concentration

5.1.1 Structure and function of chloroplasts

Consolidation of knowledge

PAGE 163 **1** Before you begin exploring this topic, it is important to be clear about the correct terminology. How you use and understand language will be important throughout this chapter. Complete the task below and ensure you are clear on the use and meaning of each term. Label the chloroplast in Figure 5.1 with the following list of terms. (Remember the rules for labelling, see page 36.)

- stroma
- outer membrane
- granum
- DNA
- thylakoid
- ribosomes
- inner membrane
- thylakoid membranes

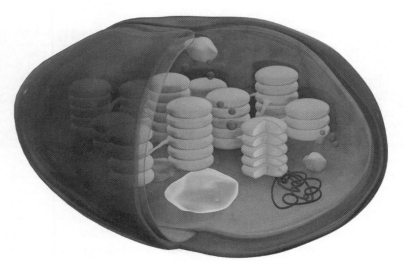

Figure 5.1 A chloroplast

2 Photosynthesis is a complex process made up of two main stages. These stages will be referred to many times. Your knowledge of each stage will help you build towards a more complex understanding of photosynthesis as a complete biochemical pathway. On the chloroplast in Figure 5.2, label the stages of photosynthesis and write the inputs and outputs of each stage in the correct positions in the white boxes.

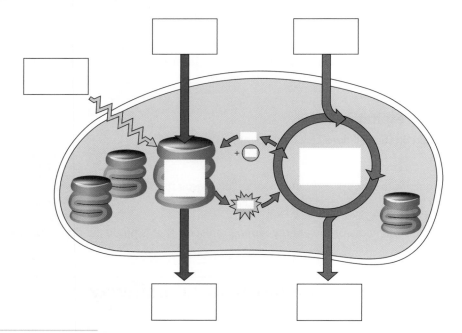

Figure 5.2 A chloroplast and the stages of photosynthesis

3 In the following photosynthetic equation, state the name of the molecule that is the source of carbon in all carbohydrates.

$$12H_2O + 6CO_2 \xrightarrow[\text{Chlorophyll}]{\text{Light}} 6O_2 + C_6H_{12}O_6 + 6H_2O$$

5.1.2 Photosynthesis in C₃, C₄ and CAM plants and the role of Rubisco

Consolidation of knowledge   PAGE 167

The Calvin cycle is a three-stage process shown in Figure 5.3a. During carbon fixation, the enzyme Rubisco helps to combine carbon dioxide with a 5-carbon acceptor molecule to make a 6-carbon compound. Rubisco then splits this 6-carbon compound into two 3-carbon compounds. Take some time to look at Figure 5.3 and make the connections between the text and the diagram as a way to answer the questions below.

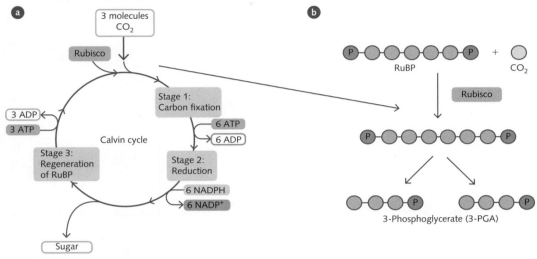

Figure 5.3 a The Calvin cycle; **b** the action of Rubisco

1   What are the three stages in the Calvin cycle?

_____

_____

_____

2   During which of these three stages is Rubisco involved?

_____

3   If Rubisco were exposed to a high level of heat or a low pH, what would be the effect on the progress of the Calvin cycle? Explain your answer.

_____

_____

4   A non-competitive inhibitor of Rubisco was added to the carbon fixation stage of the Calvin cycle. What would be the effect on the progress of the Calvin cycle? Explain your answer.

_____

_____

5   From Figure 5.3a, determine the energy needs of the three stages of the Calvin cycle. Provide evidence from the figure for each of your answers.

a   Stage 1 – carbon fixation

_____

b   Stage 2 – reduction

_____

c Stage 3 – regeneration of RuBP

_____

_____

6 Most plants are C_3 plants. The C_3 refers to the first carbon compound produced during photosynthesis (Figure 5.3b). However, some plants have evolved C_4 photosynthesis, which is an adaptation to environments of high temperature and light and low water. Contrast the functioning of C_3 plants with that of C_4 plants.

_____

_____

_____

_____

_____

_____

7 Some plants use the crassulacean acid metabolism (CAM) pathway during photosynthesis as an adaptation to living in very dry environments.

a Construct a table to compare the photosynthetic processes of C_4 and CAM plants.

b   How do the different photosynthetic pathways in C_4 and CAM plants aid their survival in their natural environments?

_____

_____

_____

_____

_____

_____

_____

_____

_____

_____

5.1.3 Inputs and outputs of photosynthesis – Part A

Key science skills

Develop aims and questions, formulate hypothesis and make predictions
- identify independent, dependent and controlled variables in controlled experiments
- predict possible outcomes

PAGE 168

Develop

Plan and conduct investigations
- determine appropriate investigations methodology: case study; classification and identification; controlled experiment; correlational study; fieldwork; literature review; modelling; product, process or system development; simulation
- design and conduct investigations; select and use methods appropriate to the investigation, including consideration of sampling technique and size, equipment and procedures, taking into account potential sources of error and uncertainty; determine the type and amount of qualitative and/or quantitative data to be generated or collated

Comply with safety and ethical guidelines
- demonstrate safe laboratory practices when planning and conducting investigations by using risk assessments that are informed by safety data sheets (SDS), and accounting for risks

Analyse and evaluate data and investigation methods
- repeat experiments to ensure findings are robust

Construct evidence-based arguments and draw conclusions
- evaluate data to determine the degree to which the evidence supports or refutes the initial prediction or hypothesis
- use reasoning to construct scientific arguments, and to draw and justify conclusions consistent with the evidence and relevant to the question under investigation
- identify, describe and explain the limitations of conclusions, including identification of further evidence required

Plants photosynthesise using the pigment chlorophyll. This pigment is found in the mesophyll cells of a leaf. Oxygen is a by-product of photosynthesis. Therefore, the buoyancy of leaves can be used as a measure of the rate of photosynthesis.

In addition to chlorophyll, carbon dioxide, water and light are required for photosynthesis. An experiment is conducted in which 10 leaf discs (Figure 5.4a) are suspended in a solution of sodium hydrogen carbonate in water. Sodium hydrogen carbonate is a source of carbon dioxide. A lamp is placed next to the leaf discs as the light source (Figure 5.4b). A stopwatch is used to record how long it takes for all the leaf discs to float to the top of the solution.

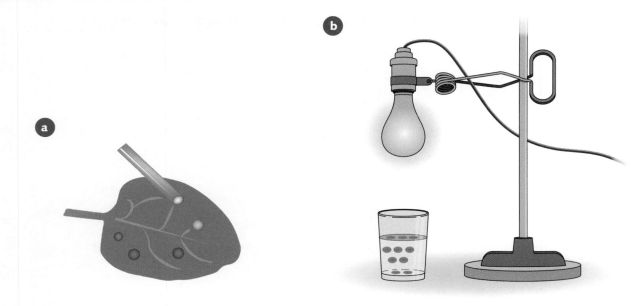

Figure 5.4 a Leaf discs are cut from a fresh leaf. **b** Ten leaf discs are placed into a solution of water and sodium hydrogen carbonate and a light is shone on them.

Complete the questions below based on your understanding of the information provided.

1   Design an experiment to test the following hypothesis.

If carbon dioxide is a key requirement of photosynthesis, then photosynthesis will stop if carbon dioxide is no longer provided.

2   a   What are the independent and dependent variables in this experiment?

   b   What are the extraneous variables, and how will each of them be controlled?

c Describe the control set-up.

d What outcome would support the hypothesis?

e What outcome would refute the hypothesis?

3 What safeguards can be built into the experimental method to ensure that the data is valid and precise?

4 What precautions can be taken to minimise the effects of systematic errors?

5 Complete a risk assessment for the experiment using Table 5.1.

Table 5.1 Risk assessment

What are the risks in doing this experiment?	How can these risks be managed to stay safe?

6 Construct a results table to record the data for your experiment.

7 The results of the experiment show that the leaf discs that did not have access to added sodium hydrogen carbonate took more time to rise to the surface than the leaf discs that did have access to sodium hydrogen carbonate. Interpret these results in terms of the progress of each stage of photosynthesis.

_____

_____

_____

_____

8 Write a conclusion for this investigation, taking the hypothesis, data collected and any limitations of the investigation into account.

_____

_____

_____

_____

_____

_____

5.1.4 Factors that affect the rate of photosynthesis

PAGE 168

Key science skills
Develop aims and questions, formulate hypothesis and make predictions
- identify independent, dependent and controlled variables in controlled experiments
- predict possible outcomes

Analyse and evaluate data and investigation methods
- process quantitative data using appropriate mathematical relationships and units, including calculations of ratios, percentages, percentage change and mean

Construct evidence-based arguments and draw conclusions
- use reasoning to construct scientific arguments, and draw and justify conclusions consistent with the evidence and relevant to the question under investigation
- discuss the implications of research findings and proposals.

Develop

Light

A Year 12 Biology class wanted to test whether different colours of light affected the rate of photosynthesis of a plant. Each of four groups of students set up six test tubes containing a similar-sized piece of the green freshwater plant _Elodea_. The students filled the test tubes with pond water so that the plants were covered. Each group was allocated a different colour of light to test: sunlight, red, blue and green. One group exposed their test tubes to sunlight. The other three groups exposed their test tubes to light from a lamp covered in cellophane of their allocated colour. To measure photosynthetic rate, the students counted the number of bubbles that each plant produced for 5 minutes. The combined class results are shown in Table 5.2.

Table 5.2 The number of oxygen bubbles observed in 5 minutes

Trial	Sunlight	Red light	Blue light	Green light
1	26	36	29	16
2	26	37	30	16
3	27	39	31	15
4	28	38	30	14
5	25	40	30	16
6	29	40	30	15
Mean				

Complete the questions provided based on the information presented.

1   Write a hypothesis for this experiment.

2   What is the independent variable in this experiment?

3   What is the dependent variable in this experiment?

4   Calculate the mean for the results for each colour of light and enter these into the final row of Table 5.2.

5   Which colour of light resulted in the highest average photosynthetic rate?

6   Which colour of light resulted in the lowest average photosynthetic rate?

7   What was the purpose of the test tubes exposed to sunlight? Explain your answer.

8   Explain the results obtained by the Biology class in terms of absorption of light by the chlorophyll.

9   Discuss one implication of this research in relation to crop production.

Temperature and carbon dioxide

Two agricultural scientists conducted an experiment to test the effect of carbon dioxide concentration and temperature on the rate of photosynthesis in wheat plants. They wanted to find out if increasing carbon dioxide concentration and temperature would lead to higher crop production. The results of the experiment are shown in Figure 5.5. Take some time to connect the information provided to the results presented in the graph in Figure 5.5. Use what you have learnt about this experiment to answer the questions below.

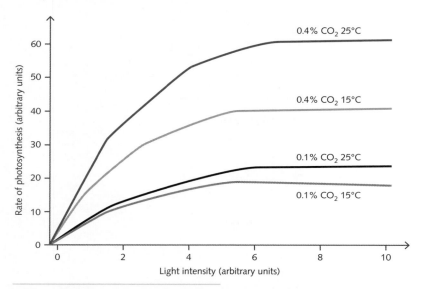

Figure 5.5 Results of the investigation to test the effect of carbon dioxide concentration and temperature on the rate of photosynthesis in wheat plants

1 Write a hypothesis for this experiment.

2 What are the independent variables in this experiment?

3 What is the dependent variable in this experiment?

4 Relate each independent variable to its effect on the relevant stage of photosynthesis and explain how it could lead to a change in the rate of photosynthesis.

5 Use appropriate biological terminology and reasoning to explain why all four graphs level out at around light intensity 6 even though light intensity continues to increase.

_____

_____

_____

_____

_____

_____

6 Do the results support or refute the hypothesis? Provide evidence from the results to support your answer.

_____

_____

_____

_____

7 State two implications of this research for the increasing temperatures around the world due to climate change.

_____

_____

_____

_____

8 Write a conclusion of no more than 60 words. In your conclusion you need to:

- restate the hypothesis
- state whether the hypothesis is supported or refuted
- state the main findings
- suggest scope for further investigation given your findings.

_____

_____

_____

_____

5.1.5 Inputs and outputs of photosynthesis – Part B

Key science skills

Analyse and evaluate data and investigation methods

- identify and analyse experimental data qualitatively, handling where appropriate concepts of: accuracy, precision, repeatability, reproducibility and validity of measurements; errors (random and systematic); and certainty in data, including effects of sample size in obtaining reliable data

Construct evidence-based arguments and draw conclusions

- evaluate data to determine the degree to which the evidence supports or refutes the initial prediction or hypothesis
- use reasoning to construct scientific arguments, and draw and justify conclusions consistent with the evidence and relevant to the question under investigation
- identify, describe and explain the limitations of conclusions, including identification of further evidence required

Develop

TB
PAGE 169

A study was undertaken by students to determine if carbon dioxide is used during photosynthesis. Phenol red is an indicator that is red in a basic solution and yellow in an acidic solution. When carbon dioxide dissolves in water, carbonic acid is produced.

The students designed an investigation with six test tubes, all of which had 20 mL of water and 5 drops of phenol red. Four of the test tubes had the water plant *Elodea* added. These test tubes were numbered 1–4. A lamp was used as a light source.

» Test tube 1 was 10 cm from the lamp.
» Test tube 2 was 20 cm from the lamp.
» Test tube 3 was 10 cm from the lamp and wrapped in aluminium foil
» Test tube 4 was 20 cm from the lamp and wrapped in aluminium foil.

Test tubes 5 and 6 did not contain *Elodea*. Test tube 6 was wrapped in aluminium foil. Both tubes were placed 30 cm from the lamp.

After 2 hours, the following results were obtained.

Test tube	Initial colour of water	Colour of water at 2 hours
1	Red	Red
2	Red	Red
3	Red	Yellow
4	Red	Yellow
5	Red	Red
6	Red	Red

Complete the questions below based on your understanding of the information provided.

1 Write a hypothesis for this investigation.

_____

_____

2 What type of data did the students collect?

_____

3 How valid and precise is this data?

_____

4 What is the purpose of test tubes 5 and 6?

_____

_____

5 From the results, could the students conclude that distance from the light source affected the amount of photosynthesis by the plants? Explain.

_____

_____

_____

6 Does the data support or refute the hypothesis? Justify your answer.

_____

_____

 9780170452618

7 Write a conclusion for this investigation, taking the hypothesis, data collected and limitations of the investigation into account.

_____

_____

_____

_____

_____

_____

_____

5.2   Cellular respiration as a biochemical pathway

Key knowledge
Cellular respiration as an example of biochemical pathways
- the main inputs, outputs and locations of glycolysis, Krebs Cycle and electron transport chain including ATP yield (details of biochemical pathway mechanisms are not required)
- the location, inputs and the difference in outputs of anaerobic fermentation in animals and yeasts
- the factors that affect the rate of cellular respiration: temperature, glucose availability and oxygen concentration

5.2.1 Cellular respiration

Key science skills
Analyse, evaluate and communicate scientific ideas
- use clear, coherent and concise expression to communicate to specific audiences and for specific purposes in appropriate scientific genres, including scientific reports and posters

Develop   TB PAGE 172

1 Figure 5.6 is missing some key labels and information. Complete this figure by adding the labels to the boxes and arrows. This will show the events that occur in cellular respiration.

    Make sure you understand this system and its process because this will help you answer Question 2. Otherwise, gather some resources and do some broader reading to prepare for giving a more detailed account of the events of cellular respiration.

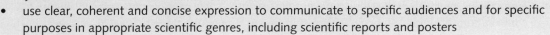

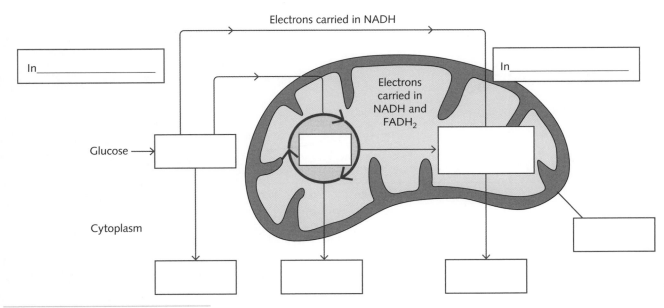

Figure 5.6 Cellular respiration

2   Using the information in Figure 5.6, use appropriate biological terminology to write a description of the events of cellular respiration in a eukaryotic cell. Include the location and all inputs and outputs (including quantities of ATP) of these events.

_____

_____

_____

_____

_____

_____

_____

_____

_____

_____

_____

_____

5.2.2 Cellular respiration using oxygen

Key science skills

Analyse, evaluate and communicate scientific ideas

*   use clear, coherent and concise expression to communicate to specific audiences and for specific purposes in appropriate scientific genres, including scientific reports and posters

Develop

Cellular respiration is a process by which all living things convert glucose into usable energy known as ATP. Figure 5.7 shows the percentage contributions of aerobic and anaerobic respiration in supplying energy to the cells of a person who is running 3000 m. Take some time to connect the information provided to the results presented in the graph. Use what you have learnt to answer the questions below.

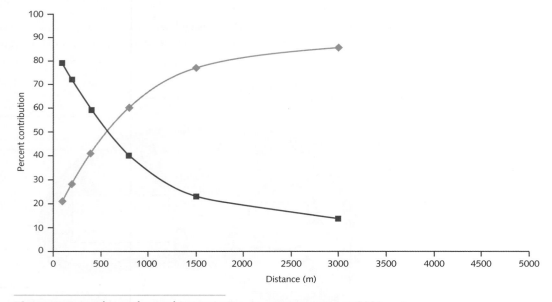

Figure 5.7 Aerobic and anerobic respiration in a person running 3000 metres

1   Label the lines on the graph to show which one represents aerobic respiration and which one represents anaerobic respiration.

2   Indicate on the graph the point where aerobic respiration took over from anaerobic respiration.

3 Extend both lines on the graph to show what would happen if the person ran a further 2000 metres.

4 Add another line on the graph to show what is happening to ATP stores.

5 Describe, using appropriate biological terminology, what is going on inside the person's cells while they are running the first 500 metres.

_____

_____

_____

_____

_____

6 Describe, using appropriate biological terminology, what is going on inside the person's cells while they are running between 1500 and 3000 metres.

_____

_____

_____

_____

_____

_____

5.2.3 Cellular respiration without oxygen

Key science skills
Analyse, evaluate and communicate scientific ideas
- discuss relevant biological information, ideas, concepts, theories and models and the connections between them

Develop

TB
PAGE 181

In the space below, construct a Venn diagram to compare aerobic respiration and anaerobic respiration. In your Venn diagram, indicate at least two similarities and five differences.

5.2.4 Putting photosynthesis and aerobic cellular respiration together

PAGE 184

Key science skills
Analyse, evaluate and communicate scientific ideas
• discuss relevant biological information, ideas, concepts, theories and models and the connections between them

Develop

Figure 5.8 shows a single cell from the multicellular green algae *Spirogyra*. Different stages in two biochemical processes that occur within the cell are shown. Take some time to study this figure to work out what it is telling you.

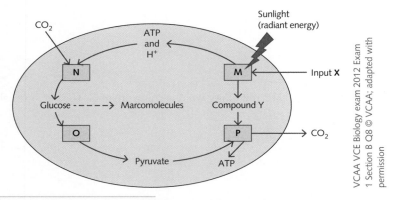

Figure 5.8 Single cell from the multicellular green algae *Spirogyra*

1 State the processes represented at:

   a M _____

   b N _____

   c O _____

   d P _____

2 What is input X?

_____

3 When grana are removed from M, compound Y is no longer produced. Describe the effect that this will have on N.

_____

_____

4 Complete Table 5.3 to show whether the removal of the molecules listed will cause an **increase**, **decrease** or **no change** in the process occurring in the part of the cell labelled M, N and P.

Table 5.3 The effects of removing molecules from biochemical processes

Removed molecule	Effect on		
	M	N	P
Oxygen			
Carbon dioxide			
Glucose			
Pyruvate			

5.3 Biotechnological applications of biochemical pathways

Key knowledge

Biotechnological applications of biochemical pathways
- potential uses and applications of CRISPR-Cas9 technologies to improve photosynthetic efficiencies and crop yields
- uses and applications of anaerobic fermentation of biomass for biofuel production.

5.3.1 Applications of CRISPR-Cas9 technologies

Key science skills

TB
PAGE 186

Construct evidence-based arguments and draw conclusions
- distinguish between opinion, anecdote and evidence, and scientific and non-scientific ideas

Analyse, evaluate and communicate scientific ideas
- analyse and evaluate bioethical issues using relevant approaches to bioethics and ethical concepts, including the influence of social, economic, legal and political factors relevant to the selected issue

Develop

The article below presents a variety of views of the gene editing technology CRISPR. Read this information with a critical eye and reflect on whether each viewpoint is opinion, anecdote or evidence. Use the information from the article to answer the questions below.

Gene editing with CRISPR

The new gene editing technology CRISPR is the talk of the scientific community. It has the potential to be an exact, fast and economical way to edit DNA. If you think about DNA as the letters on this page, then CRISPR is the pen that you can use to edit them precisely. With the pen, you can edit the letters in one word, you can change whole phrases or you can change complete paragraphs at precise locations. For example, you can edit the sentence I WANT TO MAKE A PIE by changing the letter M to B. The sentence then becomes: I WANT TO BAKE A PIE. This subtle change has changed the meaning of the sentence.

Compare CRISPR to the first-generation transgenic techniques, which were like adding a new paragraph to a book and hoping it did not ruin the story. Most plant breeders and scientists used a technique that randomly inserted DNA into a genome. This was a slow and tedious process that involved thousands of plants and hundreds of hours. DNA was inserted randomly and by good luck some landed in a useful spot. This random nature of genetic engineering meant that getting a new trait into a commercial crop and getting the product to market could take up to 10 years. Now, with CRISPR, new traits can be expressed and ready to go within two years.

So, what can be done with this amazing new technology? Scientists say that they will be able to genetically engineer foods much more easily.

CRISPR has the potential to make fruits and vegetables much healthier and more appetising. Gone will be the days of children hiding broccoli in their milk. Perhaps new varieties of fruit and vegetables can be created to join the trendy superfoods that are already on supermarket shelves. When shoppers at a local grocery store were asked if they would buy foods that had been genetically altered using CRISPR, some of the responses were: 'Certainly not, I don't want to start glowing in the dark', 'I might give it a go and see what it tastes like' and 'My sister has a friend who is a scientist and she says that it could have potential impacts on my young children, so probably not.' One person said, 'If its crisper, then it has got to be fresher right? So yes, I would try it.'

Some people do not realise that CRISPR has been known about since 1987 as a bacterial defence mechanism. Since then, scientists have been busily unlocking its secrets. In 2012, CRISPR's potential for gene editing was discovered. Jointless tomatoes, fungus-resistant bananas, non-browning mushrooms, and high-yield corn, soy and wheat have been developed using CRISPR. Scientists have used CRISPR to systematically knock out one gene at a time and see what happens to the plant without that gene. This is how a group of researchers in Mexico are approaching the development

(continued)

of a non-browning avocado, but first they must locate the browning gene to knock it out. The researchers have collected the genomes of hundreds of varieties of avocadoes and are painstakingly sifting through the DNA to find differences in the genomes of browning and non-browning types. The research is controversial because many people do not like playing with nature. Some people believe that we do not have the scientific

or regulatory ability to control unchecked evolutionary changes to species.

As the world's population grows, technologies such as CRISPR could come to be the saviour of our planet, allowing us to reduce the footprint of the agricultural industry and preserve wild ecosystems. CRISPR has the drawback of making it easier for people with bad intentions to do harm.

1   Use three different coloured highlighters, pencils or pens, to annotate the extract to distinguish opinion, anecdote and scientific ideas. Make sure you create a key to show what each colour represents.

2   Identify one bioethical issue in the use of CRISPR. Identify the approach and ethical concepts that could be used in discussing this issue. What types of external factors (social, economic, legal or political) could have an influence on this issue?

_____

_____

_____

_____

_____

_____

3   What is your opinion on the use of CRISPR technology to edit the genomes of species? Explain your answer.

_____

_____

_____

_____

5.3.2 Anaerobic fermentation of biomass for biofuel production

Consolidation of knowledge

PAGE 187 Use the information in the article below and your knowledge to answer the questions that follow.

Using biomass as a source of energy

The use of biomass and other waste materials to produce energy has the potential to mitigate global warming due to climate change. We need to stop looking at waste as something to be buried in the ground or disposed of; instead, we should look at waste as a priceless resource to be used. The world is slowly seeing the value of waste and there is a rush to use biomass and other waste materials as the raw materials for biofuel, mainly because they are readily available and cheap.

Biomass is lignocellulosic in nature and includes woodchips (Figure 5.9), sawdust, herbivorous animal waste, sugar cane, corn, shredded paper, straw, algae and much more. The energy contained in the chemical bonds of these materials could be used to run our cars and trucks, generate electricity, and produce gas and fertilisers. The need to dig up and burn fossil fuels would be gone, and with it, further climate change.

(continued)

The conversion of biomass to energy does not come at an exorbitant cost because the processes are simple. Coal-fired power stations can be replaced if biomass is used as an alternative fuel supply to heat water to produce steam to drive turbines. Either the biomass is burnt or the carbohydrates are extracted from the cells by destroying the cell walls. The extracted carbohydrates are then converted using the cells' own enzymes to ethanol by the process of fermentation.

Getty Images/Vladimir Smirnov\TASS

Figure 5.9 Woodchips, not coal, could be used to drive our electricity generators.

1   What is meant by 'biomass is lignocellulosic' in nature? (Hint: Lignocellulosic is a combination of two words – lignin and cellulose.)

_____

_____

2   Explain why the waste from carnivores such as dogs cannot be used to produce biofuel.

_____

_____

3   Explain what is meant by the following statement.

   *The energy contained in the chemical bonds of these materials could be used to run our cars
   and trucks, generate electricity, and produce gas and fertilisers.*

_____

_____

_____

4   Write the word equation for the fermentation of carbohydrates extracted from cells, showing all inputs and outputs.

5.4 Chapter review

5.4.1 Key terms

PAGE 192 The two parts of Table 5.4 are not aligned. The first part of the table presents key terms numbered 1–15. The second part presents definitions labelled A–O. Your task is to ensure that each term is matched with its correct definition by completing Question 1.

Table 5.4 Photosynthesis key terms and definitions

Number	Key term		Letter	Definition
1	C_3 plant		A	A structure in the leaf that can be opened or closed and allows for gas exchange
2	Light dependent		B	Organisms that fix CO_2 in the cytosol ahead of the Calvin cycle to improve the efficiency of photosynthesis
3	Light independent		C	Membrane structure of the organelle responsible for the light dependent reaction
4	Thylakoid		D	Most common form of plant; Rubisco binds CO_2 to a 5-carbon chain
5	Thylakoid membrane		E	A series of chemical reactions that occur during the light-independent stage of photosynthesis
6	Granum		F	A structure comprising multiple thylakoids stacked upon each other
7	Stroma		H	Green pigment that absorbs light energy
8	Rubisco		G	A stage of photosynthesis, occurring in the thylakoid, where water is split to form hydrogen ions for the next stage and O_2 as a waste product
9	C_4 plant		I	A product of fermentation in fungi and plants
10	CAM plant		J	A membrane-bound organelle that contains the green pigment chlorophyll and is found in the cytoplasm of plants and algae; its main function is photosynthesis and storage of carbohydrates
11	Calvin cycle		K	A structure found in the chloroplast that is the site of the light dependent stage of photosynthesis. These structures form the granum.
12	Chlorophyll		L	An enzyme associated with photosynthesis that helps fix carbon during the light-dependent stage
13	Ethanol		M	A plant that only fixes CO_2 at night to reduce water loss
14	Stomata		N	The liquid component of the chloroplast that is responsible for the light independent stage
15	Chloroplast		O	A stage of photosynthesis, occurring in the stroma, where carbon is fixed to form carbohydrates including glucose

1 Match the photosynthesis key terms in Table 5.4 with their definitions by writing the corresponding letter against each number in the spaces below. One has been done for you as an example.

1 _____D_____   9 _____

2 _____   10 _____

3 _____   11 _____

4 _____   12 _____

5 _____   13 _____

6 _____   14 _____

7 _____   15 _____

8 _____

2 The list of key terms below relates to cellular respiration. Use three key terms at a time to construct sentences to show that you understand the meaning of each of the key terms. Use each key term only once. An example has been done for you.

cytosol	mitochondrial matrix	aerobic respiration
cristae	Krebs cycle	ATP
mitochondria	acetyl-CoA	pyruvate
mitochondria inner membrane	electron transport chain	ATP synthase

Key terms used: anaerobic respiration, lactic acid, ethanol

Sentence: Cellular respiration in the absence of oxygen, or **anaerobic respiration** in plant cells, produces **ethanol** as a by-product, while anaerobic respiration in animal cells produces **lactic acid** as a by-product.

_____

_____

_____

_____

_____

_____

_____

_____

_____

_____

_____

_____

_____

_____

_____

Chapter review (continued)

PAGE 193

Exam
practice

5.4.2 Exam practice

1 ©VCAA 2018 Q8 (adapted) MEDIUM The diagram below shows a section through part of a mitochondrion.

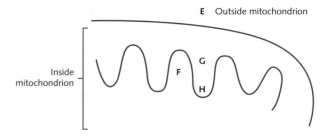

The sites of the pathways in aerobic respiration are

A E – glycolysis, F – Krebs cycle and H – electron transport chain.
B H – glycolysis, E – Krebs cycle and G – electron transport chain.
C F – glycolysis, T – Krebs cycle and E – electron transport chain.
D E – glycolysis, G – Krebs cycle and H – electron transport chain.

The following information relates to Questions 2 and 3.
The following three-dimensional diagram is of an organelle found in eukaryotic cells.

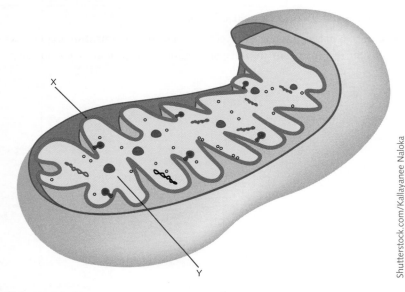

Shutterstock.com/Kallayanee Naloka

2 ©VCAA 2017 Q9 (adapted) MEDIUM The region labelled X is called the

A matrix.
B crista.
C inner membrane.
D inter-membrane space.

3 ©VCAA 2017 Q10 (adapted) HARD The structure labelled Y

A is where glucose enters glycolysis.
B provides more surface area for reactions to occur.
C is where most ATP is synthesised.
D is where pyruvate is broken down, releasing carbon dioxide.

4 ©VCAA 2017 Q11 (adapted) **HARD** A plant cell culture was exposed to radioactively labelled carbon dioxide. The cells were exposed to light and monitored for five minutes.

After this time, the radioactively labelled carbon atoms would be present in which cellular chemical?

A Adenosine triphosphate

B Oxygen

C Water

D Glucose

The following information relates to Questions 4 and 5.

The graph below shows the net output of oxygen in spinach leaves as light intensity is increased. Temperature is kept constant during the experiment.

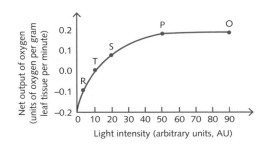

5 ©VCAA 2017 Q13 (adapted) MEDIUM Which one of the following conclusions can be made based on the graph?

A At point S, photosynthesis is no longer occurring.

B The optimal level of light intensity for photosynthesis occurs at point P.

C At point S, the amount of oxygen output is one-quarter of that at point P.

D Below 10 AU of light intensity, the aerobic respiration rate is smaller than the photosynthesis rate.

6 ©VCAA 2017 Q14 (adapted) MEDIUM The rate of oxygen output remains constant between points P and O because

A the availability of Rubisco limits the rate of photosynthesis.

B heat has denatured the enzymes involved in the photosynthesis reactions.

C the light intensity has damaged the chlorophyll molecules in the spinach chloroplasts.

D high levels of oxygen produced at point P have accumulated around the spinach leaves, resulting in no more oxygen being produced.

7 ©VCAA 2009 E1 Q3 (adapted) As a result of the light-independent stage of photosynthesis, uncharged (energy) carriers are produced.

a Name one of these uncharged carriers.   1 mark

b Availability of water is a limiting factor in photosynthesis. Name one other limiting factor.   1 mark c All algae, cyanobacteria and plants that photosynthesise contain the pigment chlorophyll a. Some contain additional pigments such as chlorophyll b.

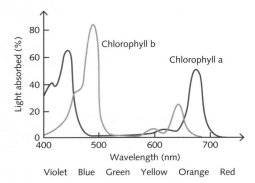

Refer to the graph above.
  i What is the purpose of additional photosynthetic pigments? Provide evidence for your answer.   2 marks

_____

_____

  ii Explain why neither chlorophyll a nor chlorophyll b absorbs light between the wavelengths of 500 and 600 nm.   2 marks

_____

_____

d Scientists exposed two groups of identical plants to a range of temperatures. One group was kept in a low light intensity environment and the other group was kept in a high light intensity environment. The following graph summarises the results obtained by the scientists.

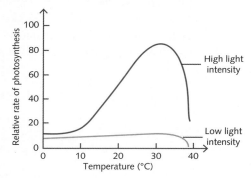

Account for the rate of photosynthesis in both groups at:
  i 10°C and below.   2 marks

_____

_____

  ii 30°C.   2 marks

_____

_____

  iii 40°C.   2 marks

_____

_____

Responding to antigens

6

Remember

TB
PAGE 202

In Unit 1 of VCE Biology, you learnt about cells and the cell cycle, and plant and animal systems. Use the questions below as prompts to reflect on this knowledge. Answer the questions, and then check what you have written by reviewing the related chapters in your textbook. Once you have done this, you will be ready to move on to this new topic.

1 What is meant by a disease?

2 List three different diseases and state which parts of the organism they affect.

3 What type of cell is a bacterium? What process does it use to reproduce?

4 State two different types of coverings found on plant leaves and state the purpose of each.

5 How do plants exchange gases through their leaves?

6.1 ## Physical, chemical and microbiota barriers in animals – first line of defence

Key knowledge

Responding to antigens
- physical, chemical and microbiota barriers as preventative mechanisms of pathogenic infection in animals and plants

---

6.1.1 Physical and chemical barriers

PAGE 204 The body's first line of defence against disease is to prevent a pathogen from entering the body. This is done in several ways. Figure 6.1 shows 10 physical and chemical barriers to pathogenic infections in humans. Label these barriers and explain how each prevents the entry of pathogens.

Consolidation of knowledge

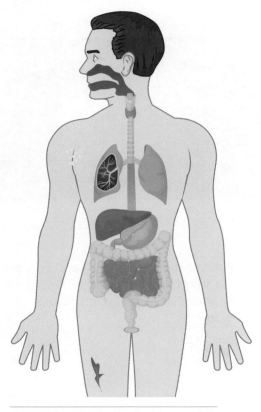

Figure 6.1 The body's first line of defence

6.1.2 Microbiota as a barrier

PAGE 206

Key science skills

Plan and conduct investigations
- design and conduct investigations; select and use methods appropriate to the investigation, including consideration of sampling technique and size, equipment and procedures, taking into account potential sources of error and uncertainty; determine the type and amount of qualitative and/or quantitative data to be generated or collated
- work independently and collaboratively as appropriate and within identified research constraints, adapting or extending processes as required and recording such modifications

Comply with safety and ethical guidelines
- demonstrate safe laboratory practices when planning and conducting investigations by using risk assessments that are informed by safety data sheets (SDS), and accounting for risks

Generate, collate and record data
- record and summarise both qualitative and quantitative data, including use of a logbook as an authentication of generated or collated data

Analyse and evaluate data and investigation methods
- repeat experiments to ensure findings are robust

Develop

Antibiotics are chemicals that help to prevent the growth of bacteria. There are many different antibiotics and many ways in which they work. Some antibiotics target the enzymes responsible for synthesising the bacterial cell wall, some inhibit the enzymes involved in protein synthesis and others damage bacterial plasma membranes.

Scientists can investigate the effectiveness of antibiotics by growing bacteria in Petri dishes containing agar and placing absorbent paper discs soaked in different antibiotics onto the agar. If a bacterial species is susceptible to an antibiotic, it will not grow in the area surrounding the disc containing that antibiotic. This area is known as the zone of inhibition. The larger the zone of inhibition, the more susceptible the bacterial species is to that antibiotic.

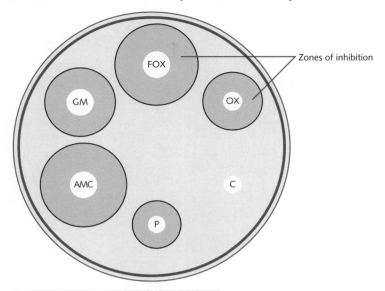

Figure 6.2 Zones of inhibition for five different types of antibiotic

Figure 6.2 shows an agar plate with six paper discs, five soaked in a different antibiotic and one control. The zones of inhibition are shown in grey.

1 Use a ruler to measure the diameters of the zones of inhibition. Record the diameter of each zone of inhibition in Table 6.1.

Table 6.1 Zone of inhibition sizes for different antibiotics

Antibiotic	Zone of inhibition (mm)
Penicillin (P)	
Amoxicillin (AMC)	
Gemifloxacin (GM)	
Cefoxitin (FOX)	
Oxacillin (OX)	
Control (C)	

2 List the antibiotics in order from most effective to least effective against the bacterium.

3 If you were to choose an antibiotic to treat an illness caused by this bacterium, which would you choose?

4 *Staphylococcus aereus* is a common bacterium in the human microflora but it can cause illness in humans. Overuse of antibiotics has led to antibiotic resistance in this species. You have been asked to investigate whether other substances could be used to control *S. aereus*. Adapt the technique above to design an investigation to find out how susceptible *S. aereus* is to a variety of household materials such as ginger, garlic juice, liquid soap, household bleach, disinfectant solution and eucalyptus oil.

When designing your investigation, make sure that you:
- state your hypothesis
- choose the most suitable methodology to test your hypothesis
- identify your dependent and independent variables
- include a control

- use a sterile technique
- identify all risks and state ways to manage them
- collect results that are valid and precise
- reduce any sources of error
- state the results you would predict if your hypothesis was supported
- create a results table to record your results.

Physical and chemical defences in plants

Key knowledge

Responding to antigens
- physical, chemical and microbiota barriers as preventative mechanisms of pathogenic infection in animals and plants

6.2.1 Plants – first line of defence

Plants, like animals, can be invaded by pathogens. Plants also have a variety of physical and chemical barriers. Figure 6.3a is a transverse section of a leaf, while Figure 6.3b is a drawing of a cactus plant. Label these figures to show the physical barriers that plants use to keep invaders out.

Consolidation of knowledge

PAGE 211

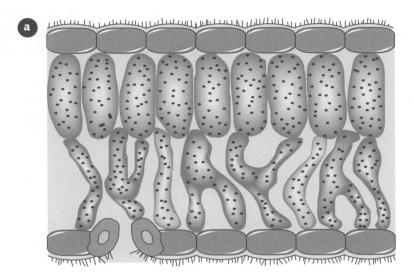

Shutterstock.com/Zonda

Figure 6.3 a A transverse section of a leaf; **b** a cactus plant (not to scale)

6.2.2 Plants – second line of defence

Key science skills

Develop aims and questions, formulate hypotheses and make predictions
- identify, research and construct aims and questions for investigation

Construct evidence-based arguments and draw conclusions
- discuss the implications of research findings and proposals

Develop

PAGE 213

Read the following article and then use the information from the article to answer the questions below.

Vaccinating tomato plants to repel pathogens

When plants come under attack from invading bacteria, viruses or fungi, they mount a two-pronged response, producing both offensive chemicals to kill invaders and defensive chemicals to prevent the invasion from spreading further across the plant. Researchers at Stanford University in the United States have developed a chemical vaccine to help plants to switch on these defence mechanisms before they are overwhelmed by an invading pathogen.

The drug is the naturally occurring defensin called *N*-hydroxypipecolic acid (NHP). NHP triggers a series of chemical responses that make uninfected leaves less appetising to pathogens. The research team found that NHP helped plants to switch on this defensive mechanism and repel pathogens earlier. NHP switches on the systemic acquired resistance response, which creates a chemical defence to protect uninfected tissues.

The economically important and popular tomato plant was used in this study and the infection was by the bacterium that causes black speck. Once successfully lodged in a tomato plant, black speck can progressively turn leaves yellow until the plant dies. The researchers investigated whether NHP triggered the systemic acquired resistance response in tomato plants as it did in other plants. The researchers administered NHP on the underside of tomato leaves. They then infected the leaves with the bacterium that causes black speck. All the treated plants remained free of the disease. Plants that had water applied to their leaves instead of NHP were overcome with the disease.

1 Explain why black speck causes the death of plants.

_____

_____

_____

2 **a** What research question were the researchers wanting to answer?

_____

_____

b What was the answer to this research question?

_____

_____

3 What methodology did the researchers use to structure their investigation?

_____

4 What was the purpose of applying water to the underside of some leaves?

_____

_____

5 What further research could the researchers undertake? (Hint: The research could be especially useful in the area of crop plants to feed the human population.)

_____

_____

_____

6 What techniques could be used as part of this further research?

6.3 Innate immune response in animals – second line of defence

Key knowledge

Responding to antigens

- the innate immune response including the steps in an inflammatory response and the characteristics and roles of macrophages, neutrophils, dendritic cells, eosinophils, natural killer cells, mast cells, complement proteins and interferons

6.3.1 Some specific cells of the innate immune response

Key science skills

Generate, collate and record data

- organise and present data in useful and meaningful ways, including schematic diagrams, flow charts, tables, bar charts and line graphs.

Reinforce

TB
PAGE 214

In humans, if a pathogen such as a bacterium breaches the first line of defence, it will be detected and dealt with by the host's immune system. The initial response to a pathogen is rapid and general and occurs in the same way every time a pathogen invades the body. This response is called the innate immune response and is sometimes described as the second line of defence.

Activity: Draw a flow chart of the innate immune response

In this activity, you will create a flow chart for the following scenario: a young boy called Tom steps on a rusty nail and the bacterium *Clostridium tetani*, which causes the disease tetanus, enters his body via the puncture wound on his foot.

You will need scissors and a glue stick.

What to do

Step 1 Develop a flow chart that depicts the response of Tom's innate immune system. Draw your flow chart on page 136.

Step 2 Add detail to your flow chart by cutting and pasting the corresponding cells in Table 6.2 (page 137) to their correct position. You may draw the cells instead; drawing is an alternative that may aid you in remembering this information.

Step 3 Annotate your flow chart with labels and descriptions to demonstrate your understanding of the processes and cells involved.

Create your flow chart here.

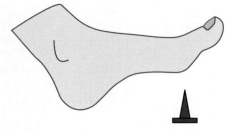

Cut out the cells in Table 6.2 to add detail to your flow chart on page 136.

Table 6.2 Cells involved in the innate immune response

Cell type	Cell shape				
Mast cell					
Macrophage					
Natural killer cell					
Dendritic cell					
Monocyte					
Neutrophil					
Basophil					
Eosinophil					

6.3.2 Innate immune response – inflammation

Jerry stood on a piece of broken glass that penetrated 1.3 cm into his foot. Figure 6.5 shows two of the three phases that Jerry's wound will undergo to heal.

Consolidation
of knowledge

PAGE 220

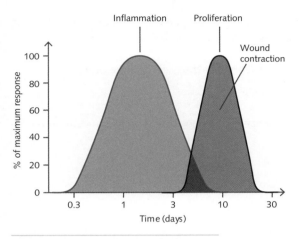

Figure 6.5 Phases of wound healing

1 Describe the cells involved in wound healing and the activities occurring at the site of the injury during the first two days.

_____

_____

_____

_____

_____

_____

_____

_____

_____

2 Describe the activities occurring at the site of the injury on the third day.

_____

_____

_____

_____

TB
PAGE 224

6.4   **Antigens and pathogens**

Key knowledge
Responding to antigens
* initiation of an immune response, including antigen presentation, the distinction between self-antigens and non-self antigens, cellular and non-cellular pathogens and allergens

Several pathogens can invade organisms and produce disease. These are either cellular pathogens (e.g. bacteria) or non-cellular pathogens (e.g. viruses).

Consolidation of knowledge

1   List two other types of cellular pathogens and one other type of non-cellular pathogen.

_____

_____

_____

2   Use the terms in the following table to label the bacterium and virus in Figure 6.6.

Bacterium	Virus
Cell wall	Core proteins
Flagellum	Viral nucleic acid
Antigen 1	Phospholipid envelope
Antigen 2	Viral proteins

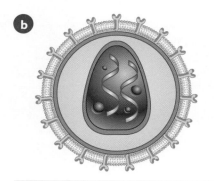

Figure 6.6 a A bacterium and **b** a virus

3 Viruses rely on live host cells for reproduction. Figure 6.7 shows a T-bacteriophage virus invading a live bacterial cell. Explain what is happening at each stage in Figure 6.7.

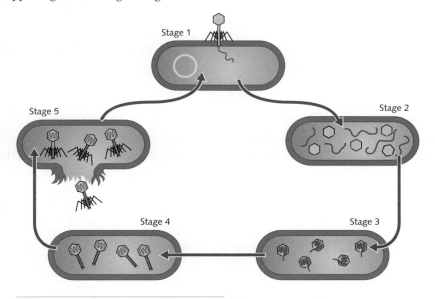

Figure 6.7 The five stages of T-bacteriophage virus invading a live bacterial cell

a Stage 1

_____

_____

_____

b Stage 2

_____

_____

_____

c Stage 3

_____

_____

_____

d Stage 4

_____

_____

_____

e Stage 5

_____

_____

_____

_____

6.5   Allergens

PAGE 231

Key knowledge
Responding to antigens
- initiation of an immune response, including antigen presentation, the distinction between self-antigens and non-self antigens, cellular and non-cellular pathogens and allergens

Key science skills
Analyse and evaluate data and investigation methods
- identify and analyse experimental data qualitatively, handling where appropriate concepts of accuracy, precision, repeatability, reproducibility and validity of measurements; errors (random and systematic); and certainty in data, including effects of sample size in obtaining reliable data

Develop

Five children were in a playground for a fifth birthday party. While running around, one of the children was stung by a bee, initiating an allergic reaction. Use your knowledge of the allergic reaction to answer the questions below.

1   What is a physical sign that someone is suffering from an allergic response?

_____

2   Describe the steps that would occur at the cellular level when someone is suffering an allergic reaction.

_____

_____

_____

_____

_____

_____

_____

3   Allergic reactions are common among people, and anaphylaxis is the most severe form of allergic reaction. Anaphylaxis is a generalised allergic reaction that often involves more than one body system (e.g. skin, respiratory, gastrointestinal and cardiovascular). Anaphylaxis can rapidly become life threatening and must be treated as an emergency requiring immediate medical attention. Figure 6.8 shows the causes of anaphylaxis admissions to hospitals in Australia by age group during the period 1994–2005. Study this graph carefully to make sure you understand what it is telling you, and use this information to answer the questions below.

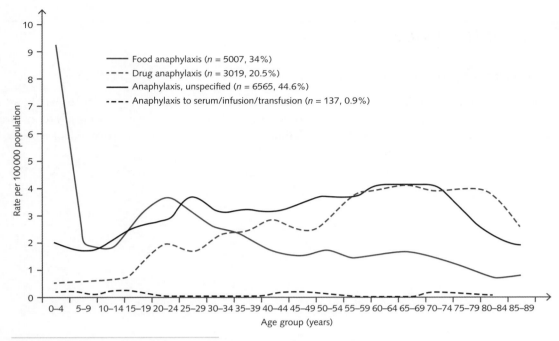

Figure 6.8 Causes of anaphylaxis admissions to hospitals in Australia by age group 1994–2005

'Reproduced with permission from The Royal Australian College of General Practitioners from: Laemmle-Ruff I, O'Hehir R, Ackland M, Tang ML. Anaphylaxis: Identification, management and prevention. Aust Fam Physician 2013;42(1–2):38–42. Available at www.racgp.org.au/afp/2013/januaryfebruary/anaphylaxis

a Which age group experiences the highest rate of food anaphylaxis?

b Which age group experiences the highest rate of drug anaphylaxis?

c Provide one reason why the rate of anaphylaxis to serum/infusion/transfusion is low compared with other causes.

d Describe the trend in the rate of unspecified cause of anaphylaxis from 0 to 89 years.

e Provide one reason for the downward trend of the drug and unspecified causes of anaphylaxis after 80 years.

f Comment on the validity and reliability of this data.

6.6　Chapter review

PAGE 239

6.6.1 Key terms – spot the errors

Spelling biological terms correctly is one component in succeeding at VCE Biology. Take the time to learn how to spell the key terms correctly. This activity is designed to assist you with this. Each of the following key terms is spelt **incorrectly**. Write the correct spelling and provide the definition for each key term in the space provided below.

allargen

_____

_____

antegen

_____

_____

bacteriphage

_____

_____

celluler pathigen

_____

_____

compliment protein

_____

_____

defencins

_____

_____

dendrytic cell

_____

_____

eosophils

_____

_____

histomine

_____

_____

inflamation

lyses

lysosyme

makrophage

monosyte

neutrofil

pathigen

phagositosis

preon

vasedilation

6.6.2 Exam marking

The following two exam questions have been answered by a VCE Biology student. Below each question is the marking guide the exam markers were provided with to correct that question. Your job is to become the exam marker and correct each answer and decide how many marks you will allocate the student's answer. Provide a final mark for each question in the space provided.

Exam
practice

TB
PAGE 241

Question

1   ©VCAA   2019 Section B Q3 (adapted)   The human immune system consists of a series of defensive barriers that protect the body from infection. When bacteria come into contact with the body, they immediately encounter these defences.

a   List one physical defence of the human body against invasion by bacteria. (1 mark)
Intact skin

b   The bacteria enter the human body through the nose. List one physical defence that the body uses to try to expel the bacteria from the body and explain its function. (2 marks)
Hairs in the nose trap the bacteria

c   If the bacteria breach the first line of defence of the human body, they will encounter the human immune system. List one of the cells that will be encountered and describe the function of that cell in relation to the bacteria. (2 marks)
eosinophil cells destroy the plasma membranes of the bacteria

Marks: /5

Marking guide

1 a   Intact skin or keratinised skin (1 mark)

  b   Hairs in the nose (1 mark) trap the bacteria in mucus and push this to the throat where it is coughed or sneezed out or swallowed (1 mark).

  c   Students can state any of the cells of the innate immune system; must be spelt correctly (1 mark); and state its correct function (1 mark); e.g. eosinophils (1 mark) secrete an enzyme that breaks down the bacterial cell wall (1 mark).

Question

2 ©VCAA 2005 E1 Section B Q2 (adapted) The following diagram shows two examples of pathogens. The relative size of each is indicated.

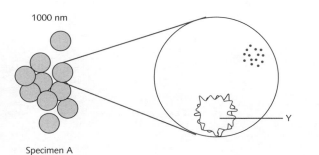

a **i** In specimen A, state what structure Y is composed of. (1 mark)
Nucleoid

ii What is the function of this structure? (1 mark)
To carry genes

b Specimen B is an example of a pathogenic agent.

i What type of pathogenic agent is specimen B? (1 mark)
Virus

ii Distinguish between a pathogenic agent and a pathogenic organism in terms of their reproduction. (1 mark)
Can't live outside a cell

Marks: /4

Marking guide

2 **a** **i** DNA or deoxyribonucleic acid (1 mark).
Must be spelt correctly.

ii The function of the DNA is to carry genetic information between generations OR to code for proteins. (1 mark)

b **i** A virus (1 mark)

ii Pathogen organism can reproduce outside a living cell whereas a pathogenic agent cannot. (1 mark)
Few students could give a significant feature that distinguishes a pathogenic agent from a pathogenic organism. Incorrect responses that gave 'smaller size' or 'becoming living within a cell' as the feature were not awarded a mark.

7 Acquiring immunity

Remember

TB
PAGE 248

In Unit 3 of VCE Biology, you learnt about the innate immune response. The questions below offer important prompts that will ensure that you have the knowledge you need to work through this chapter. Revisit the innate immune responses and answer these questions while you review the chapters as a way to help you consolidate these ideas.

1 Describe the first and second lines of defence in complex animals.

2 Differentiate between pathogens and antigens.

3 It is often said that the innate immune system provides a non-specific response to an infection. Explain this statement.

4 Differentiate between PAMPs and DAMPs.

5 What is the role of PRRs when talking about PAMPs and DAMPs?

7.1   Adaptive immune response – third line of defence

Key knowledge

Acquiring immunity

- the role of the lymphatic system in the immune response as a transport network and the role of lymph nodes as sites for antigen recognition by T and B lymphocytes
- the characteristics and roles of the components of the adaptive immune response against both extracellular and intracellular threats, including the actions of B lymphocytes and their antibodies, helper T and cytotoxic T cells

7.1.1 Lymphatic system – an analogy

Key science skills

Analyse, evaluate and communicate scientific ideas

- discuss relevant biological information, ideas, concepts, theories and models and the connections between them

Develop

PAGE 250

A bank robbery occurred at a local bank branch and a series of coordinated responses from bank staff and others removed the bank robbers from the bank before they reached the money. This scenario is similar to how the body responds to an invasion by a pathogen. The lymphatic system is made up of a network of vessels, glands, organs, nodes and specific immune cells, all with key roles to play in removing invaders.

Use Table 7.1 to show how the bank security system is analogous to the lymphatic system. The events in a bank robbery are shown in the first column. In the second column, write the features of the lymphatic system that play a similar role in the expulsion of pathogens from the body.

Table 7.1 An analogy between a bank security system and the lymphatic system

Bank security system	Lymphatic system
Bank robbers	
Bank walls	
Metal detectors (detecting the robbers' guns)	
Security cameras	
Security guards	
Police officers	
Wanted posters showing the faces of the robbers in case of future break-ins	

7.1.2 Cells of the adaptive immune system

TB
PAGE 253

Key science skills
Analyse, evaluate and communicate scientific ideas
*   discuss relevant biological information, ideas, concepts, theories and models and the connections between them

Develop

The innate immune system produces cells that react with a quick defence against invaders. These cells are shown in the left-hand circle of the Venn diagram in Figure 7.1. The innate immune system works with the adaptive immune system to fight pathogens.

Fill in the right-hand circle and the overlapping section of the Venn diagram in Figure 7.1 by drawing and naming the cell(s) that function in both the innate and adaptive immune systems and those that function in the adaptive immune system only.

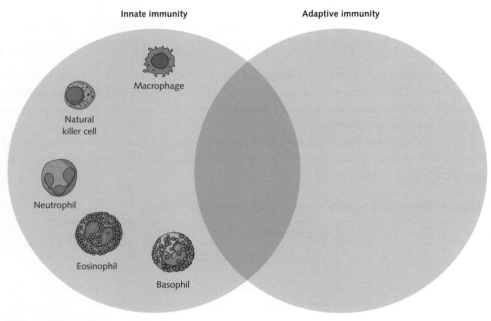

Figure 7.1 Innate and adaptive immunity

7.1.3 Intracellular or extracellular pathogens

TB
PAGE 256

Key science skills
Comply with safety and ethical guidelines
*   apply relevant occupational health and safety guidelines while undertaking practical investigations
*   demonstrate ethical conduct when undertaking and reporting investigations
Generate, collate and record data
*   organise and present data in useful and meaningful ways, including schematic diagrams, flow charts, tables, bar charts and line graphs

Reinforce

A patient went to their doctor complaining of inflamed lymph nodes in their neck, a sore throat and a runny nose. The doctor made some observations and concluded that the patient was suffering from a common cold. The patient asked for antibiotics, but the doctor said these would not be effective against a common cold.

The doctor explained to the patient that there are two broad categories of pathogens: intracellular and extracellular. A patient is prescribed an antibiotic when the pathogen is extracellular and is recognised by the body as non-self. A bacterium is an extracellular pathogen.

1   Using the term 'intracellular' in your answer, explain why the doctor did not prescribe antibiotics for the patient's cold.

_____

_____

2   The doctor wanted to do some blood tests to definitively rule out the possibility of COVID-19. When a blood sample goes to the laboratory, the laboratory technician needs to ensure that they are working to strict safety guidelines. Describe the safety guidelines that the laboratory technician would be following to ensure that the blood sample does not get contaminated.

_____

_____

_____

_____

3   When the laboratory technician analysed the blood sample, they looked for certain types of blood cells. What type of cells would the laboratory technician look for to positively diagnose COVID-19?

_____

_____

4   The laboratory technician did find an increase in this type of cells.

   a   What conclusion can now be made about the type of pathogen that is making the patient unwell?

   _____

   b   What types of cells would be found in the patient's blood sample and how are these produced to fight this type of pathogen?

   _____

   _____

5   Describe the differences between intracellular and extracellular pathogens in terms of MHC class I and MHC class II markers.

_____

_____

_____

_____

_____

6   Use a flow chart to show how the cells involved in the second line of defence become antigen-presenting cells to help with the adaptive immune system, which is known as the third line of defence.

7   Bacteria are extracellular pathogens and will initiate a cellular response, activating B leukocytes into action. A student discussed this process with his Biology teacher and stated that if the bacteria is different, the response will be the same. Create a diagram to show how clonal selection creates many clones of the lymphocyte with the correct receptor to bind to the shape of the invading antigen.

| 7.2 | **Humoral immunity** |

Key knowledge

Responding to antigens

• the characteristics and roles of the components of the adaptive immune response against both extracellular and intracellular threats, including the actions of B lymphocytes and their antibodies, helper T and cytotoxic T cells

 ## 7.2.1 Antibodies

PAGE 260 B cells produce an array of specific antibodies that tag foreign antigens for destruction.

Consolidation of knowledge

1   Label the diagram of an antibody in Figure 7.2 with the following terms.

• Variable region

• Constant region

• Antigen binding sites

• Hinge region

• Light chain

• Heavy chain

Figure 7.2 An antibody

2   Figure 7.3 is an example of an antibody-mediated immune response to a pathogen. Explain what is happening at each level of the response.

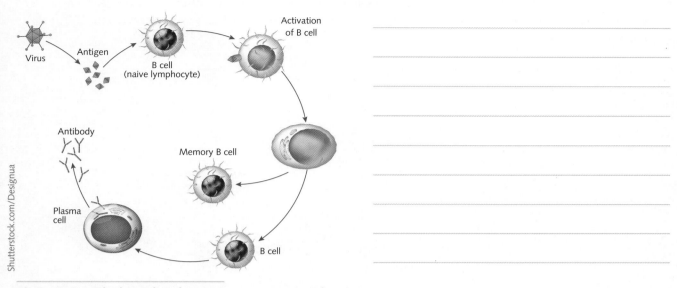

Figure 7.3 Antibody-mediated immune response to a pathogen

7.3   Cell-mediated immunity

Key knowledge
Acquiring immunity
- the characteristics and roles of the components of the adaptive immune response against both extracellular and intracellular threats, including the actions of B lymphocytes and their antibodies, helper T and cytotoxic T cells

7.3.1 Cell-mediated response

1   Figure 7.4 is an example of a cell-mediated immune response to a pathogen. Explain what is happening at each level of the response.

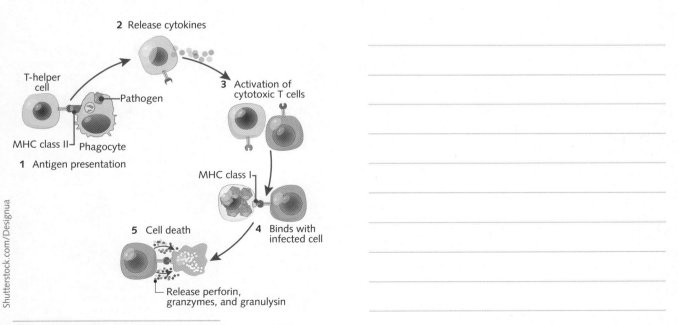

Figure 7.4 Cell-mediated immune response to a pathogen

2 In late 2019, an outbreak of a novel SARS-CoV-2 virus caused a disease known as COVID-19 in the city of Wuhan in Hubei province in China. As of mid-2020, there was no safe and effective vaccine to prevent COVID-19. Scientists developing vaccines for COVID-19 are conducting trials in animal models. To evaluate the effectiveness of a new vaccine, both humoral and cell-mediated responses are measured in the animal models. Contrast the two different types of responses.

_____

_____

_____

_____

_____

_____

_____

7.3.2 Adaptive immune responses

Key science skills
Analyse, evaluate and communicate scientific ideas
- discuss relevant biological information, ideas, concepts, theories and models and the connections between them

Develop

Figure 7.5 shows the actions and functions of the cells of the adaptive immune system. Take some time to work out what this diagram is telling you and then use your understanding of the diagram to answer the questions below.

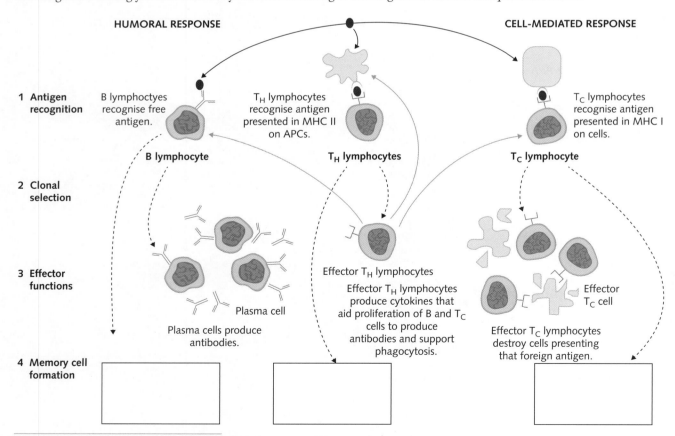

Figure 7.5 Actions and functions of the cells of the adaptive immune system

1 Compare antigen recognition for the humoral and cell-mediated responses of the adaptive immune system.

2 Does clonal selection occur in both the humoral and the cell-mediated responses of the adaptive immune system? If yes, explain how this occurs in each case.

3 Identify the effectors in both the humoral and cell-mediated responses.

4 What cells are produced from each of the responses and what are their functions? Draw the missing cells in the boxes provided in Figure 7.5.

7.4   Active and passive immunity

TB
PAGE 270

Key science skills

Develop aims and questions, formulate hypotheses and make predictions
- formulate hypotheses to focus investigation
- predict possible outcomes

Practise

1   Read the following information carefully and use it, and your own knowledge, to answer the questions below.

A small group of people were bushwalking in the Otway Ranges when one of the group members was bitten by a snake. Everyone was very concerned about the victim, but there were many different opinions about the best way to treat the bite.

The victim was not too concerned about the bite because he had been bitten by a snake and given an anti-venom before. He thought he would have immunity to this bite.

Another person did not think that this was true because anti-venom only works once and does not produce any long-term memory cells.

a   What makes an anti-venom effective against snakebites?

_____

_____

b   A third person said that, although the snakebite would need to be treated with anti-venom, it would only ever give passive immunity rather than active immunity. Explain what is meant by passive and active immunity.

_____

_____

_____

_____

c   Another person agreed that anti-venom provides passive immunity but wanted to investigate this further to determine whether it was artificial or natural immunity. This person said that it could only be natural immunity because the victim has already been bitten before.

   i   Is this statement correct? Explain your answer.

_____

_____

_____

   ii   If a series of tests was carried out on snakebite victims, what would (or would not) be found in their blood to be able to draw this conclusion?

_____

_____

_____

2   Figure 7.6 shows the response of the body to two successive vaccinations of an antigen. Five weeks after a first
     injection, a second injection was given.

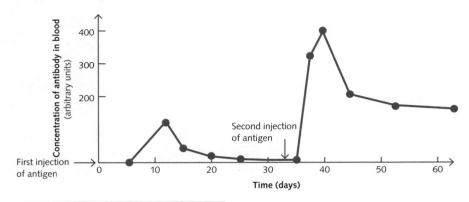

Figure 7.6 The body's response to two vaccinations of an antigen

a   Explain the peak at day 12 after the first vaccination.

_____

_____

b   Explain what is happening in the body from day 12 to day 20.

_____

_____

c   Explain what is happening in the body from day 36 to day 40.

_____

_____

d   Explain what is happening in the body from day 50 to day 60. Do you expect this line to decrease to zero? Explain
     your answer.

_____

_____

_____

_____

7.5   Chapter review

7.5.1 Key terms

PAGE 276

Key science skills

Generate, collate and record data

- organise and present data in useful and meaningful ways, including schematic diagrams, flow charts, tables, bar charts and line graphs.

Reinforce

Use the following terms to complete the flow chart in Figure 7.7 to show the functioning of the adaptive immune system. You may introduce new terms as required. You may need to use a larger piece of paper!

Lymphatic system	MHC I	Helper T cells
Bone marrow	MHC II	Neutralisation
Thymus	Antigen	Cellular response
B lymphocytes	Clonal selection	Cytokines
T lymphocytes	Antibodies	Apoptosis
B plasma cells	Agglutination	Regulatory T cells
B memory cells	Opsonisation	Memory T cells
Complement activation	Cytotoxic T cells	

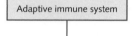

Figure 7.7 The function of the adaptive immune system

Chapter review (continued)

PAGE 277

Exam practice

7.5.2 Exam practice

The following information relates to Questions 1 and 2.

The following diagram shows a cross-section through an influenza virus.

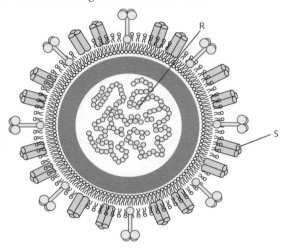

1 ©VCAA 2008 E1 Section A Q16 (adapted) EASY The part of the virus labelled R is its

A antigenic marker.

B lipid envelope.

C protein coat.

D viral genome.

2 The part of the virus labelled S

A is injected into a host cell.

B is detected by the immune system as being self.

C hijacks the functioning of the immune system.

D assists in attaching the virus to the host cell.

3 ©VCAA 2017 Section A Q26 (adapted) HARD A daily blood sample was obtained from an individual who received a single vaccination against a strain of the influenza virus. The individual had no prior exposure to this strain of influenza. The graph below shows the concentration of antibodies for this strain of influenza that were present in the individual's blood over a period of 65 days.

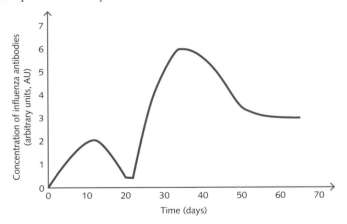

Which one of the following conclusions can be made using this data?

A Memory B cells were activated by exposure to the same strain of the influenza virus on day 12.

B B plasma cells specific to this strain of influenza were most numerous on day 35.

C Exposure to this virus in the future would not cause a response in the immune system.

D The vaccination containing weakened influenza antigens occurred on day 30.

4 ©VCAA 2005 E1 Section B Q5e–g (adapted) Antigens exist on cell surfaces. The following diagrams show the different forms of antigens, and their relative distribution, found on the surfaces of different strains of *Staphylococcus* bacteria. You wish to make a single vaccine that will be effective against as many strains as possible of the *Staphylococcus* bacteria shown.

a Which strain would you use to make the vaccine? 1 mark

_____

b Against which of the strains would your vaccine be effective? 1 mark

_____

c Students were asked to draw an antibody that would be most effective against strain P. The following diagrams were presented by three different students.

Explain which student has drawn an antibody that would be most effective against strain P *Staphylococcus* shown above. 1 mark

_____

5 ©VCAA 2010 E1 Section B Q4 (adapted) In rats, the EB12 receptor on the B cells helps determine if a cell becomes a plasma or a memory cell. Scientists used three different strains of rats to investigate B cell immunity. None of the strains had been exposed to the influenza virus. The strains were as follows.

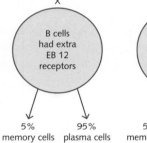 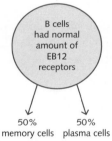 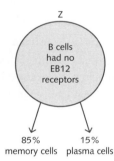

The three rat strains were infected with the influenza virus.

a Explain which strain, X, Y or Z, would be the least effective at destroying the fast-acting influenza virus.

2 marks

_____

_____

_____

_____

b Explain the actions of memory cells and plasma cells in destroying the influenza virus. 2 marks

_____

_____

_____

_____

_____

_____

_____

8 Disease challenges and strategies

Getty Images/Kulka

Remember

TB
PAGE 284

In Unit 4 of VCE Biology, you learnt about the innate and adaptive immune responses. This knowledge will help you work though this chapter of the workbook. You can now draw on your growing knowledge base about the function and structure of the immune system. This chapter focuses on the challenges presented by disease. The following questions will help focus your thoughts and identify the language that you will need to address in this area.

1 What is a disease?

2 Distinguish between an infectious disease and a non-infectious disease.

3 What causes disease?

4 Define the following terms.
   a Antigen

   b Antibody

   c Cytokines

8.1 Emerging and re-emerging pathogens

Key knowledge
Disease challenges and strategies
- the emergence of new pathogens and re-emergence of known pathogens in a globally connected world, including the impact of European arrival on Aboriginal and Torres Strait Islander peoples

8.1.1 Spread of pathogens in a globally connected world

Key science skills
Generate, collate and record data
- plot graphs involving two variables that show linear and non-linear relationships
Construct evidence-based arguments and draw conclusions
- use reasoning to construct scientific arguments, and to draw and justify conclusions consistent with the evidence and relevant to the question under investigation
Analyse, evaluate and communicate scientific ideas
- use clear, coherent and concise expression to communicate to specific audiences and for specific purposes in appropriate scientific genres, including scientific reports and posters
- acknowledge sources of information and assistance, and use standard scientific referencing conventions

PAGE 286

Develop

Epidemiologists are scientists who study diseases in populations by looking at the distributions and causes of diseases.

In 2020, the novel coronavirus SARS-CoV-2, which causes the disease COVID-19, resulted in global economic shutdowns, illness and death. Epidemiologists initally thought that the pandemic started in a wet animal market in Wuhan, China.

1   If the viral spread was thought to have started in Wuhan, China, in December 2019, explain how it had spread across the planet by March 2020.

2   Table 8.1 shows the total number of recorded COVID-19 cases over a selected period of time from four selected countries. Refer to Table 8.1 to create a graph that visually represents this information. Use the axes on page 164.

Table 8.1 Total reported number of COVID-19 cases within selected countries, February–August 2020

Country	Number of cases						
	15 Feb 2020	15 Mar 2020	15 Apr 2020	15 May 2020	15 Jun 2020	15 July 2020	15 August 2020
Australia	15	300	6447	7019	7335	10487	23035
New Zealand	0	8	1386	1498	1504	1547	1609
Sweden	1	1063	12433	30448	53132	62786	81281
Taiwan	18	59	395	440	445	451	482

Data source: https://www.worldometers.info/coronavirus/#countries

3 Refer to your graph above and discuss the general trends shown in the number of cases of each country from 15 February 2020 to 15 August 2020.

_____

_____

_____

_____

_____

_____

_____

_____

9780170452618

Table 8.2 Government response to the COVID-19 pandemic

Country	Government response
Australia	Mid-March 2020: • restrictions on social gatherings • adoption of work from home • closure of schools and adoption of remote learning • nationwide border closure and mandatory 14-day quarantine for all incoming international travellers • adoption of social distancing measures (People must be 1.5 metres apart.) • increase of testing for COVID-19 May 2020 • release of COVIDSafe app with more than 5 million downloads by 5 May.
New Zealand	Mid-March 2020: • nationwide lockdown with the aim of total virus elimination • 14-day self-quarantine for each person entering the country • strict home lockdown laws forbidding anybody leaving except for exercise nearby • testing with capacity of 8000 per day.
Sweden	• no lockdown established as other countries did • citizens advised to voluntarily practise social distancing and to work from home • adoption of 'herd immunity' strategy with the idea that most of the population would catch the virus, develop antibodies and therefore not spread the virus further.
Taiwan*	January 2020: • all people entering Taiwan from Wuhan were checked and other border controls increased • transmission routes of the virus investigated • citizens encouraged to wear masks • traced contacts of confirmed cases placed in quarantine.

*Taiwan was very badly affected by the SARS epidemic in 2003.

4 Consider the data presented in Table 8.1 and the initial response of the government of each country to the pandemic shown in Table 8.2. Evaluate the success of each country's initial response to the COVID-19 pandemic, using the information presented to you in Tables 8.1 and 8.2.

Table 8.3 Total reported number of COVID-19 cases within the United States and Brazil, February–August 2020

Country	Number of reported cases						
	15 Feb 2020	15 Mar 2020	15 Apr 2020	15 May 2020	15 Jun 2020	15 July 2020	15 August 2020
United States	15	3621	65 5452	1 494 312	2 186 553	3 622 584	5 530 201
Brazil	0	200	28610	218 223	891 556	1 970 909	3 317 832

Data source: https://www.worldometers.info/coronavirus/#countries

5   Use the graph paper below to graph the data in Table 8.3.

6   Refer to your graph in Question 5 and discuss the general trends in the number of cases for each country.

_____

_____

_____

7   Use secondary sources to research the government responses to the COVID-19 pandemic in the United States and Brazil. Using the data presented in Table 8.3 and from your research, evaluate the success of both countries' responses to the pandemic. Make sure that you correctly reference all the resources that you use.

_____

_____

_____

_____

_____

_____

_____

_____

8   Imagine that you are the Chief Health Officer of Australia in 2040. You have been asked to prepare recommendations for Australia's response to the next pandemic. List five essential recommendations that you would propose.

_____

_____

_____

_____

_____

_____

_____

_____

8.1.2 Impact of European arrival on Aboriginal and Torres Strait Islander peoples

Key science skills
Analyse, evaluate and communicate scientific ideas
• use clear, coherent and concise expression to communicate to specific audiences and for specific purposes in appropriate scientific genres, including scientific reports and posters

Develop

TB
PAGE 288

Communication is an important skill in science. Once an investigation has been completed, it is necessary to communicate the findings to others to contribute to scientific knowledge. Later, other scientists may come along and build upon your work, so it is important that your writing is clear, coherent and concise.

The following three extracts relate to the same topic but from three different points of view. Synthesise information from each extract and construct a logical and concise answer of no more than 400 words to the following question:

What effects did the arrival of Europeans in the 18th century have on the health of Aboriginal and Torres Strait Islander peoples?

Extract 1

The First Fleet arrived in Sydney in 1788, bringing with it up to 1500 convicts, marines, seamen, civil officers and free people. More ships arrived in the following months and years, bringing British settlers, supplies and more convicts. Also on these ships was a host of infectious pathogens that had not previously been present in the land now known as New South Wales.

These pathogens caused diseases that devastated the Indigenous populations. Diseases that were common among the early European settlers were measles, smallpox, influenza, scarlet fever, chicken pox, bronchitis and the common cold, and they were deadly for Indigenous people. Respiratory and sexually transmitted diseases spread rapidly, as the lands and waters used by Aboriginal people became inundated with European settlers and animals carrying these diseases. With no immunity developed from previous exposure to these diseases, local populations were decimated.

Within 14 months of settlement, Captain Arthur Phillip estimated that 50% of the local Indigenous population had been affected. The Wurundjeri people of the Yarra region, which is now modern Melbourne, were severely affected by the new diseases, which caused about 60% of the deaths of Aboriginal people across the Port Phillip area. Aboriginal clans around the Melbourne area felt the severe impact of European colonisation even before Europeans reached Melbourne. An earlier smallpox epidemic had spread south from Sydney and killed up to a third of the population of the eastern Australian tribes.

from *VICscience VCE Biology Units 3 & 4 4e* Chapter 8, page 288

Extract 2

I know straight away that the arrival of the First Fleet that came into our country that was the beginning of the end for the Aboriginal way of life, how it used to be. They brought with them disease, all sorts of diseases. There was hundreds and hundreds and hundreds that died of these diseases when they landed within [Botany Bay] Sydney, and then sailing down to here [Melbourne], this is where it was a direct impact on my people ... you started to lose things because of the government, this was the last chance of a traditional lifestyle for the Aboriginal people. But back then you learned at an early age, how to hunt and what to eat when you were little. Since first contact in our country, things had changed because there were settlers on our country, with cattle and so forth. Our people were not allowed to go on their traditional hunting ground because a lot of it was fenced off. So, they were trying to survive and try to find things to eat. And then in the end the government had to give them food, they had to stand in line to get a cup of flour, some rations, so that they could feed themselves and their children.

And it's like you're losing the essential ancient knowledge in history ... like the cures, the treatments that are actually in advance of the Western medicine that we have now. I guess in my generation all medicines were practice with our people and are within my family. There are certain plants that we used, like old man weed. It had magical qualities when used correctly. [Old man weed, *Centipeda cunninghamii*, is now known as the wonder weed. Its indigenous habitat was along the Murray River. It has been found to have antimicrobial, antifungal, anti-inflammatory and antioxidant properties. It is now used to treat many skin conditions and is part of many anti-ageing formulations.]

This is what my grandmother told my mother. She passed away at real old age. But she had to live to see this happening, you know, with children being taken away and all my Aunties and all that happened that people read about that was what she lived through, and then my mother and her sisters and the next generation. They made sure that it was passed on all that history just by talking about it, you know, so that's how it sort of is in enough of all those old stories from my family's oral history.

Aunty Zeta Thomson, Victorian Wurundjeri and Yorta Yorta Aboriginal Elder

Extract 3

While the innate immune response is able to prevent or control some infections, it is limited in the ways in which it can react. The adaptive immune response, which includes both B-cell-based humoral immunity and T-cell-based cellular immunity, reacts much more specifically and powerfully to invading pathogens. The steps necessary to act against viruses, such as the coronavirus that causes COVID-19, is the antibody response. Antibody responses are the main way in which vaccines protect us from infection by a variety of viruses, and the absence of protective antibodies contributes to the rapid spread of new viruses in previously unexposed and unvaccinated populations.

adapted from Harvard Medical School
https://onlinelearning.hms.harvard.edu/hmx/immunity

Write your synthesis of the three extracts here. Remember, you have been asked to offer a careful and concise response that is no more than 400 words in length.

What effects did the arrival of Europeans in the 18th century have on the health of Aboriginal and Torres Strait Islander peoples?

8.2 Strategies for controlling pathogen transmission

Key knowledge

Disease challenges and strategies

- scientific and social strategies employed to identify and control the spread of pathogens, including identification of the pathogen and host, modes of transmission and measures to control transmission

Key science skills

Analyse, evaluate and communicate scientific ideas

- critically evaluate and interpret a range of scientific and media texts (including journal articles, mass media communications and opinions in the public domain), processes, claims and conclusions related to biology by considering the quality of available evidence

Develop

Figure 8.1 A poster to encourage behaviours to stop the spread of the novel coronavirus SARS-CoV-2 that causes COVID-19

Figure 8.1 shows a poster that was used to educate the public about behaviours that would assist in stopping the spread of COVID-19 during 2020 and 2021. Refer to Figure 8.1 and provide a biological reason for each of the recommended steps to follow to prevent catching and spreading SARS-CoV-2.

8.3   Vaccination programs

Key knowledge
Disease challenges and strategies
- vaccination programs and their role in maintaining herd immunity for a specific disease in a human population

8.3.1 Herd immunity

Key science skills
Construct evidence-based arguments and draw conclusions
- evaluate data to determine the degree to which the evidence supports the aim of the investigation, and make recommendations, as appropriate, for modifying or extending the investigation
- evaluate data to determine the degree to which the evidence supports or refutes the initial prediction or hypothesis

Analyse, evaluate and communicate scientific ideas
- discuss relevant biological information, ideas, concepts, theories and models and the connections between them

Figure 8.2 shows the fatality rate (percentage of people who die once they contract the disease) against the contagiousness (ease of transmission) of the disease. Contagiousness is expressed as reproduction number or R_0 (R nought). R_0 represents the number of people an infected person transmits the disease to.

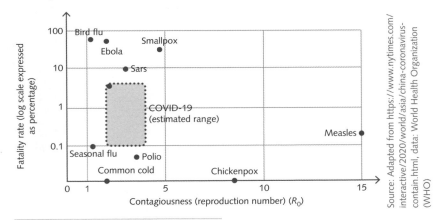

Figure 8.2 Fatality and contagiousness of infectious diseases

1   An epidemiologist predicted that if there were to be a global outbreak of measles, it would be deadlier than the common cold. Use Figure 8.2 to provide evidence to support or refute this prediction.

_____

_____

_____

_____

2 Define 'contagiousness' and describe the relationship between contagiousness and fatality. How are these two concepts shown in Figure 8.2?

_____

_____

_____

_____

3 Compare the contagiousness of Ebola to that of measles.

_____

_____

_____

_____

4 In September 2019, there was an outbreak of measles in Samoa in which there were more than 5700 cases and 83 deaths. A vaccine for measles has been available since the 1970s and has been very successful in reducing the number of cases around the world. Epidemiologists discovered that before the outbreak, the rate of immunisation for the measles virus in the population of Samoa had dropped to 34%.

a Use your knowledge of herd immunity to account for the outbreak of measles in Samoa in 2019.

_____

_____

_____

b During the measles outbreak, schools were closed and Christmas celebrations were cancelled to try to prevent the spread of measles. The measles virus is highly virulent and the mode of transmission is direct contact with an infected individual, airborne droplets or indirect contact (although this is not as common). What was the significance of these restrictions in terms of control measures against the transmission of measles?

_____

_____

_____

_____

c Describe other measures that could have been adopted during the measles outbreak to try to control its spread.

_____

_____

_____

_____

_____

_____

8.3.2 Vaccines

Key science skills
Analyse, evaluate and communicate scientific ideas
- analyse and evaluate bioethical issues using relevant approaches to bioethics and ethical concepts, including the influence of social, economic, legal and political factors relevant to the selected issue
- use clear, coherent and concise expression to communicate to specific audiences and for specific purposes in appropriate scientific genres, including scientific reports and posters

PAGE 304

Develop

1 Table 8.4 shows the important times in a young child's life when they will need to be immunised. A number of diseases are targeted to ensure healthy outcomes for each child and the community.

Table 8.4 Vaccination schedule for Victorian children

Age/School year level	Disease to be vaccinated against
Birth	Hepatitis B
2 months	Diphtheria, tetanus, pertussis, hepatitis B, poliomyelitis, *Haemophilus influenzae* type b, pneumococcal, rotavirus
4 months	Diphtheria, tetanus, pertussis, hepatitis B, poliomyelitis, *Haemphilus influenzae* type b, pneumococcal, rotavirus
6 months	Diphtheria, tetanus, pertussis, hepatitis B, poliomyelitis, *Haemophilus influenzae* type b, pneumococcal, rotavirus
12 months	Measles, mumps, rubella, pneumococcal, meningococcal C
18 months	Measles, mumps, rubella, diphtheria, tetanus, pertussis, *Haemophilus influenzae* type b
4 years	Diphtheria, tetanus, pertussis, poliomyelitis
Year 7	Diphtheria, tetanus, pertussis, human papillomavirus
Year 10	Meningococcal C

a Explain why the first major immunisation does not occur until two months of age. How is a newborn baby protected against these diseases prior to immunisation?

b Is immunisation an example of natural or artificial immunity? Explain your answer.

c Is immunisation an example of active or passive immunity? Explain your answer.

d Justify the need to immunise a baby at 2, 4 and 6 months with the same vaccines.

2   Figure 8.3 shows the number of deaths in Australia by decade (1926–2005) for five diseases. The arrows indicate when vaccines were introduced for the disease.

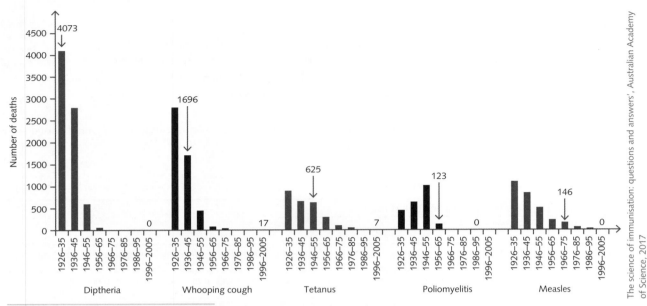

Figure 8.3 Number of deaths per decade in Australia for diphtheria, whooping cough, tetanus, poliomyelitis and measles

a   Not everyone can be immunised because they might have health issues such as allergies. If everyone is not immunised, provide reasons why deaths for all of the diseases shown have decreased dramatically over time.

b   Use the information in Figure 8.3 to develop an argument why all children should be vaccinated against these diseases.

3   Read the following article and use the information to answer the questions below.

Human challenge trials

In normal vaccine trials, thousands of participants are given a trial vaccine or a placebo and are monitored over several years to see if the vaccine prevents development of the disease. These types of trials can take up to 10 years before a vaccine is cleared for use with the general public. Due to the urgency of the need for a COVID-19 vaccine, human challenge trials have been suggested.

In human challenge trials, participants are administered a trial vaccine and then deliberately exposed to the infection. This is similar to what Jenner did in the 1790s when he exposed a young boy to cowpox. The advantage of this is that results can be available within weeks rather than years because participants are guaranteed to be exposed to the disease. To minimise the risk to participants, it has been suggested that participants should be aged 20–45 years and recruited from areas where the likelihood of infection is already high. Speeding up the development of a vaccine for COVID-19 in such a way could save hundreds of thousands of lives.

a What is the bioethical issue in the scenario in the article?

_____

_____

b Which ethical approach would you choose to resolve the bioethical issue identified in part a?

_____

_____

c Explore the scenario above in terms of the relevant ethical approach and concepts. (See page 12.)

_____

_____

_____

_____

_____

_____

8.4   Immunotherapy strategies

Key knowledge
Disease challenges and strategies
• the development of immunotherapy strategies, including the use of monoclonal antibodies for the treatment of autoimmune diseases and cancer.

8.4.1 Monoclonal antibodies and the treatment of cancer

Monoclonal antibodies are engineered molecules that produce identical copies of the same antibody. Figure 8.4 summarises the process of producing monoclonal antibodies. Use the information contained in this figure, and your own knowledge, to answer the questions below.

Consolidation of knowledge

TB
PAGE 306

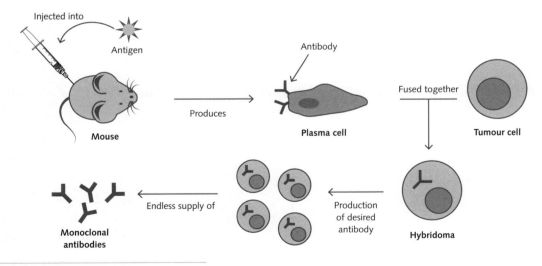

Figure 8.4 Production of monoclonal antibodies

1   What are the steps required to produce a hybridoma?

2   What are the unique properties of the hybridoma?

3   Why are these antibodies called 'monoclonal'?

4   Draw a final step to the process above to show how monoclonal antibodies work to destroy cancer cells.

5   Outline two ways that monoclonal antibodies are an improvement on the usual treatments for cancer, such as chemotherapy.

8.4.2 Cancer and autoimmune diseases

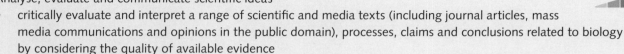

Key science skills

Construct evidence-based arguments and draw conclusions
- discuss the implications of research findings and proposals

Analyse, evaluate and communicate scientific ideas
- critically evaluate and interpret a range of scientific and media texts (including journal articles, mass media communications and opinions in the public domain), processes, claims and conclusions related to biology by considering the quality of available evidence
- acknowledge sources of information and assistance, and use standard scientific referencing conventions

PAGE 309

Develop

The following article addresses ethical issues that relate to clinical trials using monoclonal antibodies as a treatment for leukaemia and autoimmune diseases. Read this article carefully and answer the questions below.

First-in-human clinical trial

The Expert Scientific Group (ESG) was set up following the very serious adverse reactions that occurred in the first-in-human clinical trial of TGN1412 in March 2006. The trial was performed in a private clinical research unit at Northwick Park Hospital in London. TGN1412 is a monoclonal antibody that was being developed as a medicine to treat leukaemia and autoimmune diseases such as rheumatoid arthritis. In the clinical trial, six healthy male volunteers experienced severe systemic adverse reactions soon after receiving TGN1412 intravenously. All six volunteers developed a cytokine release syndrome with multi-organ failure and required intensive treatment and supportive measures that were provided by the Intensive Therapy Unit at Northwick Park Hospital. Previously, first-in-human clinical trials had a very good safety record and, as far as we can determine, the TGN1412 trial outcome, where all recipients experienced such a severe and similar adverse reaction, is unprecedented. Following these events at Northwick Park Hospital, the ESG was set up by the Secretary of State for Health, with the following terms of reference.

1 To consider what may be necessary in the transition from pre-clinical to first-in-human phase I studies, and in the design of these trials, with specific reference to:
- biological molecules with novel mechanisms of action
- new agents with a highly species-specific action
- new drugs directed towards immune system targets.

2 To provide advice in the form of a report to the Secretary of State for Health for the future authorisation of such trials with an interim report to be provided within three months.

The ESG was asked to seek the views of major stakeholders before compiling its report.

Stakeholders raised several areas of concern that were not within our terms of reference. These included topics such as the process of informed consent and clarity of information, communication between clinical investigators and clinical trial subjects before and during a trial, insurance cover, the role of Research Ethics Committees, and clinical follow-up of trial subjects who had experienced an adverse reaction. Although beyond our remit, we considered these wider concerns to be extremely important, and recommend that they should be taken up as a high priority.

Source: 'Expert Scientifc Group on Phase One Clinical Trials', HMSO, The National Archives, UK © Crown Copyright 2006, Open Government Licence v3.0

1 What was the aim of the investigation?

_____

2 What methodology was used in the investigation?

_____

3 What improvement could you suggest to the researchers about the use of this methodology?

_____

_____

4 What method was used to achieve their aim?

_____

_____

_____

_____

5 Evaluate the success of the trial for the drug TGN1412.

_____

_____

6 When scientists are developing a new drug, they conduct a series of trials before the drug can then be tested on humans. Based on this information and the excerpt above, what considerations would be required to reduce adverse reactions?

_____

_____

_____

_____

7 Find another two articles that discuss the use of the drug TGN1412 for the treatment of leukaemia and autoimmune diseases. Acknowledge the source of these articles by using standard scientific referencing conventions.

_____

_____

_____

_____

_____

_____

_____

8.5 Chapter review

8.5.1 Exam practice

PAGE 317

Exam practice

1 ©VCAA   2018 Section A Q23 (adapted) MEDIUM   Rift Valley fever is a viral disease that was first described in Kenya's Rift Valley. It is spread to people by infected mosquitoes. A person bitten by an infected mosquito should be given an injection of specific antibodies.

Following the injection, this person will have

A   artificial passive immunity.
B   artificial active immunity.
C   natural passive immunity.
D   natural active immunity.

2 ©VCAA   2018 Section A Q24 (adapted) MEDIUM   Monoclonal antibodies can be produced and used to treat different types of cancers.

Which one of the following is a correct statement about monoclonal antibodies?

A   Monoclonal antibodies are lipid molecules.
B   Monoclonal antibodies pass through the plasma membrane of a cancer cell and attach to a viral antigen in the cell.
C   Monoclonal antibodies produced from the same clone of a cell are specific to the same antigen.
D   Monoclonal antibodies produced to treat liver cancer will be identical to monoclonal antibodies produced to treat brain cancer.

The following information relates to Questions 3 and 4.

Table 8.5 compares how eight diseases spread and the number of people likely to be infected by one other infected person.

Table 8.5 Spread of some diseases and number of people likely to be affected

Disease	Transmission	Number of other people infected by one person
Measles	Airborne droplets	12–18
Whooping cough	Airborne droplets	12–17
Rubella	Airborne droplets	6–7
Polio	Faecal–oral route	5–7
Smallpox	Airborne droplets	5–7
Mumps	Airborne droplets	4–7
Severe acute respiratory syndrome	Airborne droplets	2–4
Ebola	Body fluids	1–4

3 ©VCAA   2018 Section A Q32 (adapted) MEDIUM   What would be the most effective method of preventing the spread of polio during an outbreak?

A   Wash hands thoroughly after going to the toilet.
B   Establish a needle-exchange program.
C   Vaccinate all infected people.
D   Isolate all infected people.

4 ©VCAA 2018 Section A Q33 (adapted) EASY Based on the information in Table 8.5, which one of the following statements is correct?

  A Mumps is the most contagious disease.
  B Whooping cough has a higher infection rate than smallpox.
  C More people would die from whooping cough than any other disease shown.
  D Airborne droplets are the least effective means of spreading pathogens.

5 ©VCAA 2017 Section A Q25 (adapted) HARD A recent study tested for the presence of antibodies to *Helicobacter pylori*, the bacterium that causes stomach ulcers, in the blood of 550 multiple sclerosis patients and 299 healthy people. The two groups of people had the same gender balance and were of similar ages. Exposure to *H. pylori* usually occurs by the age of two years. The results of the antibody testing showed that the rate of *H. pylori* infection was 30% lower in the women with multiple sclerosis than in the healthy women or healthy men.

The findings of this study are consistent with the suggestion that

  A monoclonal antibodies could be used to treat multiple sclerosis.
  B in females, multiple sclerosis protects against stomach ulcers.
  C stomach ulcers are one of the symptoms of multiple sclerosis.
  D in females, childhood exposure to *H. pylori* helps to protect against multiple sclerosis.

6 ©VCAA 2019 Section B Q9 (adapted) Malaria is a viral disease that is prevalent in tropical regions and kills more than one million people each year. It is most commonly transferred from one person to another by the *Anopheles* species of mosquito. The symptoms of malaria are shaking chills, high fever, headache, nausea and vomiting.

  a One way that diseases such as malaria are thought to occur is when a pathogen infects humans from an animal host. Identify **one** social factor that could lead to this transfer between hosts. 1 mark

  b When scientists attempt to identify a disease, they can look for specific antibodies in infected humans. Explain why making a correct identification of a viral pathogen is important in the control of a disease. 2 marks

  c *Anopheles* mosquitoes are not found on every continent and they cannot fly long distances. The only available malaria vaccine is Mosquirix, which requires four injections and does not produce strong immunity. Describe **three** different approaches, other than vaccination, that health officials could use to reduce the spread of malaria. 3 marks

Genetic changes in populations over time

9

Remember

TB
PAGE 328

In Unit 2 of VCE Biology, you learnt about chromosomes, genomes, genotypes, phenotypes and patterns of inheritance. In Unit 3 of VCE Biology, you learnt about the immune system and disease transmission. There are links to be made between the knowledge indicated here and the new knowledge presented throughout this chapter. Answer the questions below to determine what you know already.

1 Define the following terms.

   a Chromosome

---

   b Polypeptide

---

   c Plasmid

---

   d Pathogenic

---

2 How are the following pairs of terms related?

   a Genotype and phenotype

---

   b Homozygous and heterozygous

---

c Gene and allele

_____

_____

3 Explain complementary base pairing.

_____

_____

4 What is meant by gene expression?

_____

_____

_____

5 Explain the connections between the processes of transcription and translation.

_____

_____

_____

_____

6 Explain the roles of antigens, memory B and T cells in vaccination.

_____

_____

_____

_____

7 What is herd immunity and what does it achieve in a population?

_____

_____

_____

_____

9.1 Mutations: the source of new alleles

Key knowledge
Genetic changes in a population over time
- causes of changing allele frequencies in a population's gene pool, including environmental selection pressures, genetic drift and gene flow; and mutations as the source of new alleles

9.1.1 Point mutations

PAGE 330 Figure 9.1 shows the effect of a mutation that occurs during mitosis (a) compared to the effect of a mutation that occurs during meiosis (b). Look carefully at this figure and interpret what it is telling you. Use this information to answer Question 1.

Consolidation of knowledge

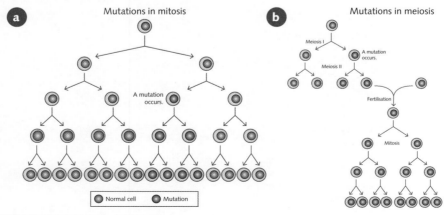

Figure 9.1 Mutations in a mitosis and b meiosis

1   Using the information in Figure 9.1 and your knowledge of cell division, explain why some mutations can be passed on to the next generation, whereas other mutations cannot.

_____

_____

_____

_____

2   Figure 9.2a shows the results of normal transcription and translation occurring inside a cell. Figures 9.2b–e show different types of point mutations occurring in DNA and the effect the mutation has on the translated protein. Label Figures 9.2b–e with the type of mutation and describe the effect this mutation has on the translated protein.

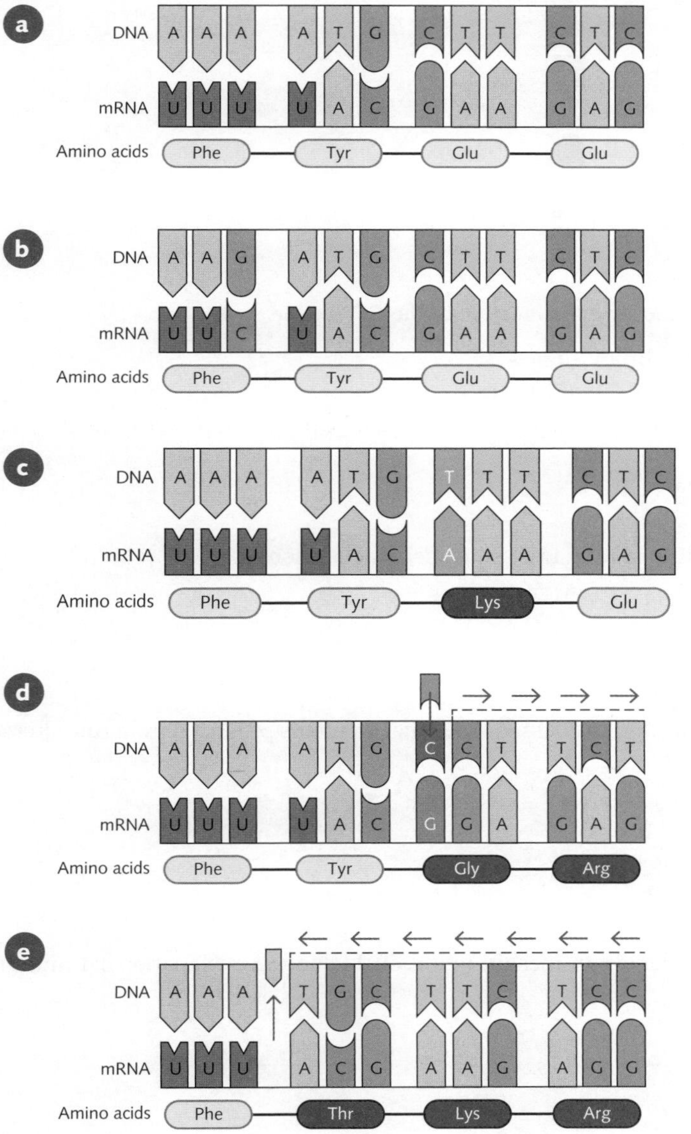

Figure 9.2 a Normal transcription and translation b–e effect of point mutations during transcription and translation

9.1.2 Cystic fibrosis mutation

PAGE 331 Cystic fibrosis affects about one in 3700 people. It is a disease of the exocrine system (the system that produces saliva, sweat, mucus and tears) that causes, among other symptoms, excessive fluid in the lungs. This eventually leads to respiratory failure. Cystic fibrosis occurs when there is a deletion mutation of just three nucleotides from the CFTR gene on chromosome 7 (see Figure 9.3). Cystic fibrosis is a recessive disease, and about one in 25 people are carriers of the mutation.

Consolidation of knowledge

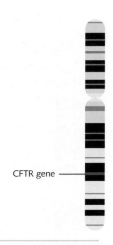

CFTR gene ——

Figure 9.3 Chromosome 7 showing the location of the CFTR gene

A normal sequence of nucleotides at this position on chromosome 7 is:

5′–ATC ATC TTT GGT GTT–3′.

1   Use the genetic code on page 41 to determine the amino acid sequence that would be produced from this nucleotide sequence.

_____

2   Cystic fibrosis occurs when three nucleotides are deleted from this sequence, which becomes:

5′–ATC ATT GGT GTT–3′

a   Which three nucleotides have been deleted?

_____

b   Use the genetic code on page 41 to determine the amino acid sequence that would be produced from this altered nucleotide sequence.

_____

c   Which amino acid has been deleted from the protein?

_____

3   Explain why the mutation in the CFTR gene is considered a frameshift mutation. Include an explanation of frameshift mutation in your answer.

_____

_____

_____

_____

9.2 Chromosomal rearrangements

Key knowledge
Genetic changes in a population over time
- causes of changing allele frequencies in a population's gene pool, including environmental selection pressures, genetic drift and gene flow; and mutations as the source of new alleles

PAGE 333

Activity: illustrating block mutations

Figure 9.4 A normal chromosome

In this activity, you will create chromosomes that illustrate four different types of block mutations. You will needs scissors and a glue stick.

What to do

Step 1 Cut out the material in Figure 9.6 on page 187 and use this to illustrate each type of block mutation listed below.

Step 2 Carefully glue the material in the spaces provided below.

Step 3 Annotate each chromosome to explain what has occurred to mutate it from the normal chromosome shown in Figure 9.4.

a Deletion

b Inversion

c Translocation

d Duplication

Bacterial conjugation

Figure 9.5 shows a process that occurs in bacteria called bacterial conjugation or horizontal gene transfer.

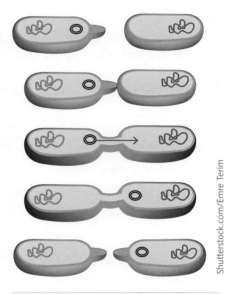

Figure 9.5 Bacterial conjugation

1 a On Figure 9.5, label the chromosomal DNA, the plasmid and the conjugation tube.

b Differentiate between horizontal gene transfer and vertical gene transfer in bacteria.

_____

_____

_____

_____

_____

_____

c Bacteria can reproduce by binary fission every 20 minutes. How would horizontal gene transfer affect the rate of evolution in bacteria?

_____

_____

_____

_____

Figure 9.6 provides you with the material that you will need to complete the modelling activity on page 185.

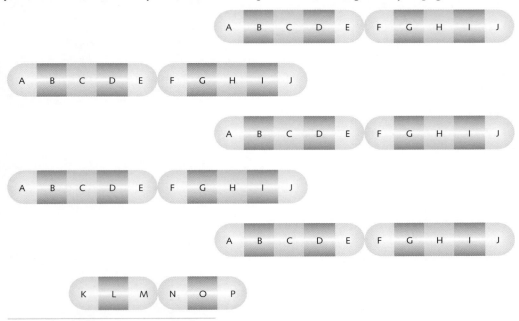

Figure 9.6 Materials for illustrating mutations

9.3 Changing allele frequencies in populations

Key knowledge
Genetic changes in a population over time
- causes of changing allele frequencies in a population's gene pool, including environmental selection pressures, genetic drift and gene flow; and mutations as the source of new alleles

9.3.1 Fowler's toad and the American toad

Key science skills
Develop aims and questions, formulate hypotheses and make predictions
- formulate hypotheses to focus investigation
- predict possible outcomes

Practise
PAGE 337

Fowler's toad (*Bufo fowleri*) and the American toad (*Bufo americanus*) are closely related species that have the ability to interbreed to produce an American × Fowler's hybrid toad. The toads both live in the flood plains of North America and have similar diets. The regions that make up their habitats are becoming increasingly encroached upon by housing estates that are taking over the flood plains. The American toad's mating season starts at the beginning of summer, while the Fowler's toad's mating season starts in late summer. Figure 9.7 graphs the populations of Fowler's toad and American toad against two factors. The nature of each factor is not relevant to this activity.

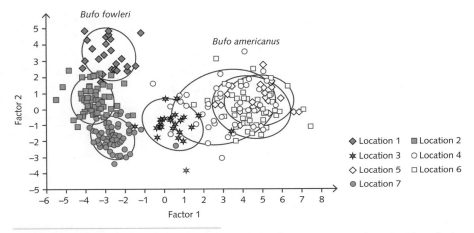

Figure 9.7 Fowler's toad and American toad populations measured against two factors

1   Which location on Figure 9.7 represents the population of American × Fowler's hybrid toads?

2   Does gene flow exist between the toads in Location 1 and the toads in Location 6? Explain your answer using evidence from Figure 9.7.

3   a   Scientists found that Fowler's toads in Location 1 had less genetic variation than toads in other locations. Propose one explanation for this observation.

b   Propose a methodology to test your explanation.

4   During an earthquake, a deep canyon was formed between populations in Locations 3 and 4. How would this affect the genetic diversity of the populations of toads on each side of the canyon.

9.3.2 Bottleneck effect

PAGE 341

> **Key science skills**
> Construct evidence-based arguments and draw conclusions
> • use reasoning to construct scientific arguments, and to draw and justify conclusions consistent with the evidence and relevant to the question under investigation
> Analyse, evaluate and communicate scientific ideas
> • use clear, coherent and concise expression to communicate to specific audiences and for specific purposes in appropriate scientific genres, including scientific reports and posters

Develop

The following article addresses some important issues that highlight the effect of bushfires and their impact on various species of native animals. There is a focus on koala populations and the pressures they face beyond this disaster. Read this article carefully and answer the questions in a clear and concise manner, making reference to the text to support your answers.

A koala bottleneck

Unprecedented bushfires throughout New South Wales and Victoria over the 2019–2020 summer resulted in millions of hectares of native bush being burnt. Unfortunately, many of the native animals that called the native bush home were also burnt. One of these native animals is the koala (*Phascolarctos cinereus*). It is estimated that thousands of koalas died in the bushfires, but a final number will never be known. Usually, a population has a large amount of genetic variation and if there are small changes in the environment, some individuals will have the capacity to survive (because they are younger, bigger, stronger, faster or smarter). In a catastrophe such as a bushfire, there is no discrimination between koalas; the event is so overwhelming that it claims the old and the young, and the weak and the strong. The koalas left behind have not been selected because of their ability to survive; instead, they happened to be in the right place at the

right time. The surviving koalas will interbreed but their offspring will not include any of the genetic variation that died out with the burnt koalas. Some of this lost genetic variation could have been advantageous to the population in the future. Removal of genetic variation from the gene pool is called a bottleneck and leads to a population that could be less genetically fit. A bottleneck makes extinction much more likely.

The Port Macquarie region in New South Wales is home to 1000–2000 koalas. If 350 koalas were to be removed from this population due to bushfires, inbreeding would increase by 20–50%. If there are no further bushfires, the population would take 5–10 years to recover. Studies of the koalas on Kangaroo Island (South Australia) and French Island (Victoria) have confirmed low genetic diversity. Koalas on the mainland were being ravaged by the disease chlamydia and healthy koalas were taken to these islands to found

(Continued)

disease-free populations. Unfortunately, there is a 19% incidence of a testicular abnormality in the male koalas on Kangaroo Island. The condition does not appear to be related to recent inbreeding but could be the result of the accidental selection of founder individuals carrying alleles for testicular abnormalities. Interbreeding amongst this isolated population has led to an increase in the frequency of the allele.

1 Define 'bottleneck' and provide an example.

_____

_____

_____

2 Explain why a bushfire is the type of catastrophic event that could lead to a bottleneck.

_____

_____

_____

3 Define 'founder effect' and provide an example.

_____

_____

_____

Table 9.1 Genetic differences between populations of koalas from 20 locations around Australia

Location	1	2	3	4	5	6	7	8	9	10	11	12	13	14	15	16	17	18	19	20
1																				
2	0.09																			
3	0.21	0																		
4	0.57	0.78	0.75																	
5	0.72	0.84	0.87	0.94																
6	0.87	0.94	0.91	1.0	0.66															
7	0.84	0.93	0.90	1.0	0.62	0														
8	0.81	0.89	0.88	0.93	0.39	0	0													
9	0.75	0.88	0.88	1.0	0.44	0	0	0												
10	0.92	0.96	0.94	1.0	0.98	1.0	1	0.97	1											
11	0.89	0.92	0.91	0.92	0.93	0.94	0.94	0.93	0.93	0.20										
12	0.74	0.88	0.88	1	0.94	1.0	1.0	0.93	1.0	1.0	0.63									
13	0.80	0.88	0.88	0.92	0.91	0.95	0.95	0.92	0.93	0.75	0.58	0.55								
14	0.89	0.94	0.92	0.98	0.97	0.99	0.99	0.96	0.99	0.92	0.60	0.87	0.19							
15	0.86	0.94	0.91	1.0	0.97	1.0	1.0	0.96	1.0	1.0	1.0	0.59	1.0	0						
16	0.85	0.93	0.91	1.0	0.97	1.0	1.0	0.96	1.0	0	0.16	1.0	0.65	0.09	1.0					
17	0.90	0.95	0.93	0.99	0.98	1.0	1.0	0.96	1.0	0.97	0.61	0.95	0.25	0	0	0.96				
18	0.78	0.85	0.86	0.89	0.88	0.92	0.90	0.90	0.65	0.58	0.46	0	0.19	0.18	0.55	0.50	0.22			
19	0.86	0.94	0.91	1.0	0.97	1.0	1.0	0.96	1.0	1.0	0.59	1.0	0.20	0	0	0.10	0	0.18		
20	0.83	0.90	0.89	0.94	0.93	0.96	0.96	0.93	0.95	0.76	0.56	0.57	0.09	0.01	0.2	0.67	0.04	0.09	0.02	

Note: Genetic difference is measured as 0 (no difference) to 1 (maximum difference)

Source: adapted from Neaves LE, Frankham GJ, Dennison S et al., 'Phylogeography of the koala, (Phascolarctos cinereus),and harmonising data to inform conservation', PLoS One 2016; 11(9): e0162207, Table 1

Location key

1	Whitsunday, Qld	11	Port Macquarie, NSW
2	Blair Athol, Qld	12	Maitland, NSW
3	Clermont, Qld	13	Campbelltown, NSW
4	Maryborough, Qld	14	East Gippsland, Vic.
5	Redlands, Qld	15	French Island, Vic.
6	Coomera, Qld	16	Cape Otway, Vic.
7	Tyagarah, NSW	17	Bessiebelle, Vic.
8	Ballina, NSW	18	Mt Lofty Ranges, SA
9	Iluka, NSW	19	Eyre Peninsula, SA
10	Pine Creek, NSW	20	Kangaroo Island, SA

4   Table 9.1 shows the genetic differences between populations of koalas from different locations around Australia. The genotypes of the different populations were compared for a series of single nucleotide polymorphisms, which are changes in a nucleotide that are present in at least 0.5% of individuals.

     a   Give one example where genetic diversity between populations is high.

_____

     b   Identify whether the Kangaroo Island population has a low or high genetic diversity. Explain your reasoning.

_____

_____

     c   From which mainland population would you say the Kangaroo Island population originally came? Explain your reasoning.

_____

_____

     d   Discuss the effect that gene flow could have on the koala founder population on Kangaroo Island.

_____

_____

_____

     e   Propose one way that the genetic diversity of the Australian koala population could be maintained.

_____

_____

_____

_____

Natural selection

Key knowledge
Genetic changes in a population over time
* biological consequences of changing allele frequencies in terms of increased and decreased genetic diversity

9.4.1 Selection pressures

Consolidation
of knowledge

TB
PAGE 343

Imagine a population of black snakes that are adapted to only being active during the night. The major predator of the snake, the mountain fox, has very good eyesight but is only able to find and kill about one snake every night due to the snakes' excellent camouflage in dark bushes. The black snake population was steadily increasing in size. In Figure 9.8, there are 14 snakes; draw a circle around each snake that you can see.

Figure 9.8 Camouflaged black snakes

Many generations ago, a mutation during meiosis in the reproductive cells of one of the black snakes created a new recessive allele that was expressed in its homozygous form as white skin. This new recessive allele was passed onto offspring but was not expressed until two snakes heterozygous for this allele mated and produced some offspring with white skin. Figure 9.9 shows the 14 snakes that resulted from this mating; draw a circle around each snake that you can now see.

Figure 9.9 Offspring from two carriers of the recessive allele for white skin

1   Identify the selection pressure in the black snake example.

_____

2   How was the new allele introduced into the black snake gene pool?

_____

3   Which snakes will the mountain fox be more likely to catch and eat?

_____

4   What will happen to the recessive white-skin allele over 10 generations if the environment remains unchanged?

_____

5   The area in which the snakes lived was developed for housing, and bright streetlights meant that black snakes could no longer hide. Figure 9.10 shows a sample of the snake population in this new environment.

   a   Identify the abiotic change that has occurred in the snake's environment.

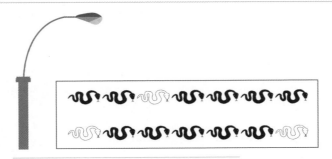

Figure 9.10 A sample of snakes from an area newly developed for housing

b Which snakes will the mountain fox now be more likely to catch and eat?

c What will happen to the recessive white-skin allele over 10 generations if this new environment remains unchanged?

9.4.2 Principles of natural selection

TB
PAGE 343

Key science skills
Analyse, evaluate and communicate scientific ideas
- discuss relevant biological information, ideas, concepts, theories and models and the connections between them

Develop

The following article provides an example of natural selection in action. Read it carefully and answer the questions below.

Natural selection in action

The Turks and Caicos Islands are a small tropical archipelago in the Caribbean Sea. The Turks and Caicos anole (*Anolis scriptus*) is a small brown lizard that is not found anywhere else in the world. Little is known about the lizard's behaviour, diet, detailed physical appearance or habitat preference.

Scientists from Harvard University (USA) and the National Museum of Natural History in France visited the Turks and Caicos Islands to find out more about this lizard. After one week spent observing, measuring and getting to know the lizard, the scientists left the island. Four days later, Category 5 Hurricane Irma hit the island and was followed a few weeks later by Hurricane Maria.

After the hurricanes, the scientists returned to the Turks and Caicos Islands to check on the surviving lizard population and to take the same measurements that

they did on their previous visit. The results surprised the scientists. They found that the surviving lizards had significantly shorter back legs, longer arms with larger toe pads and a stronger grip than the lizards they measured before the hurricanes.

To work out what selection advantage these traits conferred on the surviving lizards, the scientists set up a makeshift laboratory on the island and filmed some lizards clinging to a branch while being blown with a very strong leaf blower. This showed that lizards with longer back legs caught the wind like a sail, which caused them to lose their grip and be blown off the branch. Lizards with shorter back legs were able to tuck their legs in under their bodies and present less surface area to the wind. The result was that all four of their legs were able to stay connected to the branch.

Shutterstock.com/Emily Geraghty

Figure 9.11 A lizard from Turks and Caicos Islands

1 Use information from the article above to explain each of the following principles of natural selection.

a Individuals differ from one another; that is, individuals within populations show variation.

b Many of these variations are caused by mutations that create new alleles. All the alleles are heritable.

c In general, more offspring are born than can survive to maturity and reproduce. Because of this, there is a struggle for existence and only some organisms survive to reproduce.

d Some individuals have traits that make them more suited to their environment than other individuals. Individuals with a selective advantage are better able to reproduce and pass their alleles to the next generation.

9.4.3 Extinction

Key science skills
Analyse, evaluate and communicate scientific ideas
- analyse and evaluate bioethical issues using relevant approaches to bioethics and ethical concepts, including the influence of social, economic, legal and political factors relevant to the selected issue.

Develop

TB
PAGE 344

Below are two extracts from articles. Read the extracts and use the information contained in them, and your own knowledge, to answer the questions below.

The last remaining Bois dentelle trees

To earn a place on the world's 100 most threatened species list, you would have to be pretty rare. The *Pennantia baylisiana* tree is probably the rarest tree on Earth, with just one known specimen growing in the wild. Next on the list would have to be the *Bois dentelle*, or *Elaeocarpus bojeri*, trees, with only two known specimens growing in the wild. Both species are photoautotrophs, which means that they get their energy through photosynthesis.

Bois dentelle trees are known for their sprays of long white bell-shaped flowers and grow in the cloud forest on the island of Mauritius. The *Bois dentelle* tree has come close to extinction, not because it is commercially viable, but because it is not. The tree is being removed so that more commercially attractive exotic species such as guava can be grown. The last two *Bois dentelle* trees would have disappeared forever, taking with them the secrets locked in their genetics if it had not been for human intervention.

Adapted from Extinction watch: There are just 2 Bois Dentelle trees in the world, The Economic Times, May 12, 2020

Nature's pharmacy

Indigenous tribes have survived for centuries by using nature's pharmacy. One hundred and eighteen of the top 150 prescription drugs in the United States are based on natural sources. In 1960, a child suffering from leukaemia had a 10% chance of remission; in 1997, that had increased to 95%. The reason? Two drugs that were developed from derivatives of the rosy periwinkle, a plant that is native to Madagascar. The anti-cancer drug Taxol was first isolated from the Pacific yew tree, and anti-viral drugs that could save the world from another influenza epidemic are derived from star anise.

Adapted from Medicinal Plants at Risk, Nature's Pharmacy, Our Treasure Chest: Why We Must Conserve Our Natural Heritage, A Native Plant Conservation Campaign Report by Emily Roberson, March 2008

1   Define 'species extinction' in terms of genes and genetic diversity.

2   Identify the issue in the two articles above.

3   Identify any relevant social, economic, political and legal factors that need to be taken into consideration for the:

    a   continued removal of native species so land can be used to grow economically important crops.

    b   protection of native species so they can be studied and tested for potential life-saving medicines.

Key knowledge
Genetic changes in a population over time
•   manipulation of gene pools through selective breeding programs

9.5.1 Artificial selection: animal and plant breeding

Key science skills
Analyse, evaluate and communicate scientific ideas
•   analyse and evaluate bioethical issues using relevant approaches to bioethics and ethical concepts, including the influence of social, economic, legal and political factors relevant to the selected issue.

Develop

1   Bananas have been selectively bred over many years to produce a fruit that sells well commercially. The bananas that you see on the supermarket shelves look and taste very different from the wild type. Selective breeding has greatly reduced the genetic diversity of bananas and has created a scenario known as monoculture.

    There is an ongoing debate about the bioethics of selective breeding. Using banana monoculture as an example, highlight the ethical issues from a social, economic, legal and political point of view of selective breeding in Table 9.2.

Table 9.2 Ethical issues of the selective breeding of bananas

Social	
Economic	
Legal	
Political	

2   In the early 1800s, English bulldogs were bred for their aggression to help with controlling livestock. As a result, the skull shapes of English bulldogs changed dramatically in a short time, a clear indication of selective breeding rather than natural selection.

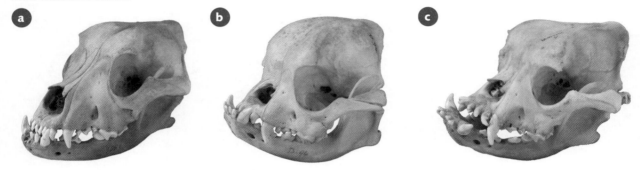

(a, b) age-fotostock/Alamy/Natural History Museum, London (c) Alamy Stock Photo/The Natural History Museum

Figure 9.12 Skulls of English bulldogs showing the effects of selective breeding: **a** the original English bulldog with a functional skull in 1860, **b** 1867 and **c** 1906, showing a very exaggerated skull. The skull has changed dramatically in a short time.

a   On Figure 9.12, label the regions on the skulls that would be helpful for controlling livestock.

b   On the same skulls, label regions that may be unhelpful for the survival of the breed.

c   Using the four steps outlined as the process for natural selection on pages 194–5 (Question 1a–d), explain how breeders of English bulldogs in the 1800s manipulated the process of natural selection for their own purposes.

3   Purebred dogs that have had traits intentionally bred into the breed are very popular. Below are some genes and their associated traits that can be selected for to produce desirable features in dogs.

- · IGFR1: A common recessive variant in this gene can result in a small dog, but only if two copies are present.
- · FGF4: A variant in this gene has the dominant effect of shortening leg length. Any dog with one copy of the variant, from either its mother or its father, will have short legs.
- · FOXI3: A variant in this gene causes hairlessness and affects tooth development. This variant is commonly seen in hairless Mexican, Peruvian and Chinese crested dogs. Hairless dogs have one copy of the mutant FOXI3 gene and one copy of the normal gene. Dogs with two copies of the mutant gene do not survive. If two hairless dogs are bred, some of the offspring will inherit two normal genes and will have normal hair. In Chinese crested dogs these are called powder puffs.

Discuss the ethics of selective breeding in purebred dogs. Decide which bioethical approach to apply and which ethical concepts to use to explore this issue. Include the influence of relevant social and economic factors.

9.6   Natural selection and consequences for disease

Key knowledge

Genetic changes in a population over time
- consequences of bacterial resistance and viral antigenic drift and shift in terms of ongoing challenges for treatment strategies and vaccination against pathogens

9.6.1 Natural selection explains antibiotic resistance

TB
PAGE 349

Key science skills

Analyse and evaluate data and investigation methods
- identify outliers, and contradictory or provisional data

Construct evidence-based arguments and draw conclusions
- evaluate data to determine the degree to which the evidence supports or refutes the initial prediction or hypothesis

Develop

1   Figure 9.13 provides two maps of Europe. Each map uses the same light grey to dark grey shading to represent low to high values for the relevant measure. Map (a) shows antibiotic use across Europe and map (b) shows the clinical resistance in the bacterium *E. coli* across Europe. Interpret the information carefully and use this to answer the questions below.

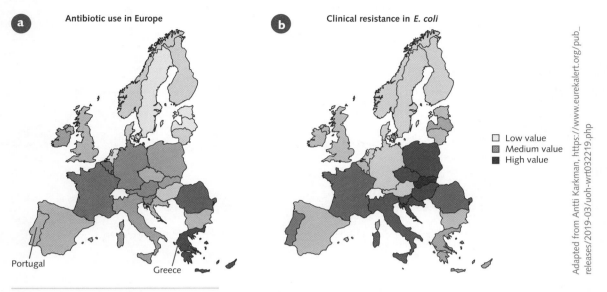

Adapted from Antti Karkman, https://www.eurekalert.org/pub_releases/2019-03/uoh-wrt032219.php

Figure 9.13 a Antibiotic use and b clinical resistance in *E. coli* in Europe

a   Analyse the correlation between antibiotic use and clinical resistance in *E. coli* in Europe.

b   Portugal has reported one of the highest resistances in *E. coli*, but its antibiotic use is in the lower part of the range. Greece has reported the highest use of antibiotics but its resistance in *E. coli* is in the lower–middle part of the range. Both of these examples do not seem to follow the general trend seen in Europe and could be considered outliers. Provide possible reasons for these outliers.

2   Ongoing research is showing that with increased antibiotic use around the world, bacteria are becoming more resistant to antibiotics.

Complete the diagram in Figure 9.14 to show how a strain of bacteria in the first Petri dish could become resistant to an antibiotic over several generations. Use the four principles of natural selection on pages 194–5 to assist you in writing a commentary to explain what is occurring at each stage of Figure 9.14.

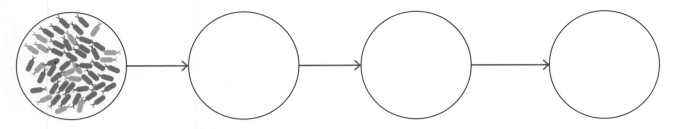

Figure 9.14

9.6.2 Viral antigenic drift and shift

TB
PAGE 356

Key science skills
Analyse, evaluate and communicate scientific ideas
• use appropriate biological terminology, representations and conventions, including standard abbreviations, graphing conventions and units of measurement

Develop

1   Use Figures 9.15a and b to differentiate between viral antigenic drift and viral antigenic shift. Make sure you include the role of the adaptive immune system and use appropriate biological terminology.

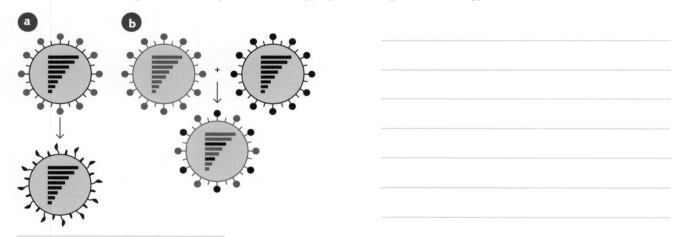

Figure 9.15 a Viral antigenic drift and **b** viral antigen shift

2   Decide whether the following occur because of viral antigenic drift or viral antigenic shift.

a   An annual flu shot is needed for the influenza virus.

b   COVID-19 is originally thought to have originated when an infected bat bit a pangolin (a scaly-skinned anteater), causing a pandemic.

9.7   Chapter review

9.7.1 Key terms

PAGE 363 Match the key term to its definition by placing the correct letter against the key term number in the answer box below.

Number	Key term	Letter	Definition
1	Allele	A	A mutation (insertion or deletion) that results in a change of all codons downstream of the mutation
2	Artificial selection	B	A group of organisms that reproduce within that isolated group
3	Population	C	The transfer of a gene from one population to another
4	Gene pool	D	The ability of a bacteria to survive the effects of an antibiotic and then reproduce to pass on that ability
5	Frameshift mutation	E	A reduction in population leading to a drastic reduction in genetic diversity
6	Natural selection	F	A sudden change to a virus's genetic material, altering the surface receptors. This may be the result of gene sharing when an individual is infected with two strains of influenza at the same time.
7	Selection pressure	G	Selection that is determined by the agricultural or cultural requirements of humans rather than natural pressures
8	Genetic drift	H	A random change in the genetic material of an organism that can be passed on to further generations
9	Gene flow	I	The survival and reproduction of certain individuals who are best adapted to the conditions in their natural environment
10	Bottleneck effect	J	An important prerequisite of selection, the range of alleles within a given population
11	Founder effect	K	A variant of a specific gene that allows for a variety of phenotypes within a population
12	Mutations	L	A small number of individuals are geographically isolated from the rest of their population and start a new population
13	Antigenic shift	M	A biotic or abiotic factor that affects the survival of particular organisms within an environment
14	Antibiotic resistance	N	The range of alleles found within a group of interbreeding individuals within a population
15	Variation	O	Random events that alter the gene frequencies in a population

Key term	Definition
1	
2	
3	
4	
5	
6	
7	
8	
9	
10	
11	
12	
13	
14	
15	

Chapter review (continued)

9.7.2 Exam practice

PAGE 365

Exam practice

1 ©VCAA 2018 Section B Q7 (adapted) Populations of the orange-spotted day gecko (a small lizard) are found on the many islands of Hawaii. There is natural variation in the phenotypes of individuals in each population.

a Explain the source of natural variation between individuals within each gecko population. 3 marks

_____

_____

_____

_____

b In 2006, a flood killed all populations of the orange-spotted day gecko on eight of the smaller islands. Scientists randomly chose eight males and eight females from a remaining population on a larger island. They introduced one male and one female to each of the eight smaller islands. Over the next four years, the scientists noted that the size of the populations had increased on each of the eight smaller islands. The scientists measured the genetic diversity in each population and found there was lower genetic diversity in each new population than in the population on the large island.

i Explain why there was higher genetic diversity in the orange-spotted day gecko population on the large island than on the smaller islands. 2 marks

_____

_____

_____

_____

ii The scientists noted that after four years, there was a significant increase in the size of the sticky toepads on the front legs of the geckos living on the smaller islands compared with those on the large island. Explain what may have happened on the smaller islands to produce this increase in the size of the sticky toepads of the geckos. 2 marks

_____

_____

_____

_____

2 ©VCAA 2013 Section B Q9 The eastern barred bandicoot (*Perameles gunnii*) is threatened with extinction in Victoria. Since European settlement, it has suffered from hunting for its fur, clearing of habitat and predation by foxes.

a What event would have to happen for the eastern barred bandicoot to become extinct? 1 mark

_____

b In 2002, seven breeding pairs of eastern barred bandicoots were bred in captivity in zoos and nature reserves in Victoria. As a result of this breeding program, 16 eastern barred bandicoots were released into a wildlife reserve near Hamilton in late 2010. Care was taken to ensure that the gene pool of the released eastern barred bandicoots was as diverse as possible.

i What is meant by 'gene pool of the released eastern barred bandicoot'? 1 mark

ii Using your knowledge of natural selection, explain what type of gene pool would be most advantageous amongst the released eastern barred bandicoots. 2 marks

c Wildlife officers are hoping that the Hamilton population of eastern barred bandicoots will increase to at least 50 in five years and that it will be maintained over time.

It was discovered that one of the breeding pairs suffered from a metabolic bone disease that caused them to become lame and unable to feed themselves. This disease was inherited, meaning that one or more of the released populations could suffer from founder effect. A small population of eastern barred bandicoots could also be affected by a genetic bottleneck effect.

Explain the meaning of each of these terms with reference to allele frequencies. 2 marks

Term	Explanation
Founder effect	
Bottleneck effect	

d The Hamilton eastern barred bandicoots will be closely monitored by surveillance cameras and radio collars after their release. Suggest one reason for this type of surveillance. 1 mark

10 Changes in species over time

Shutterstock/Lightspring

Remember

PAGE 372

In Unit 4 of VCE Biology, you learnt about how a gene pool changes over time because of environmental selection pressures, genetic drift and gene flow. You also learnt that new alleles are introduced into a gene pool through mutations and that this genetic diversity is the raw material of evolution. This knowledge will help you work through this chapter of the workbook. Test yourself by answering these questions from memory or complete this section to solidify your knowledge as you work through the chapter.

1 How do new alleles arise in a gene pool?

_____

_____

2 Describe the effect a point mutation has during transcription and translation and the resultant product.

_____

_____

_____

3 State the four principles of natural selection.

_____

_____

_____

_____

_____

4 List three factors that cause species to become extinct.

_____

_____

_____

10.1 Studying fossils

Key knowledge

Changes in species over time
- changes in species over geological time as evidenced from the fossil record: faunal (fossil) succession, index and transitional fossils, relative and absolute dating of fossils

10.1.1 Fossilisation process

Key science skills

Generate, collate and record data
- organise and present data in useful and meaningful ways, including schematic diagrams, flow charts, tables, bar charts and line graphs

Reinforce

TB
PAGE 373

Shutterstock.com/mark higgins

Figure 10.1 A fossilised fish

Figure 10.1 shows a photo of a fossilised fish. In the space below, use a technique of your choice (e.g. annotated diagrams or a flow chart) to show the steps in the fossilisation process that would have occurred to form the fossilised fish shown in Figure 10.1. You are now approaching the end of your VCE Biology course, so make sure you provide clear, concise and exact details of every step of the fossilisation process.

10.1.2 Relative dating techniques

TB
PAGE 375

Key science skills
Construct evidence-based arguments and draw conclusions
• use reasoning to construct scientific arguments, and to draw and justify conclusions consistent with the evidence and relevant to the question under investigation

Develop

Relative dating techniques are used to estimate the ages of sedimentary rocks and the fossils within them. Stratigraphic correlation is one technique used for relative dating. It compares the order in which the strata (layers) of sedimentary rocks have been laid down and is used to work out the relative ages of any fossils found in them. Examine Figure 10.2, which shows strata containing fossils from three different areas.

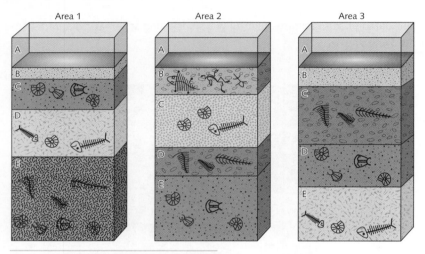

Figure 10.2 Strata of sedimentary rock from three different areas

1 If the strata remain in the order in which they were deposited, which is the oldest layer? State the area and the layer. Provide reasons for your answer.

_____

_____

_____

_____

2 List the strata from oldest to youngest, showing area and layer.

_____

_____

_____

3 Draw the fossil from Figure 10.2 in the space below that could be used as an index fossil. Provide reasons for your choice.

_____

_____

_____

_____

_____

_____

4 What are some of the assumptions being made when using this relative dating technique?

_____

_____

_____

_____

10.1.3 Absolute dating

Key science skills
Construct evidence-based arguments and draw conclusions
* use reasoning to construct scientific arguments, and to draw and justify conclusions consistent with the evidence and relevant to the question under investigation

Develop

TB
PAGE 377

Absolute dating is a technique that assigns a numerical age in years to a fossil or rock. The most common method is radiometric dating, which is based on the predictable rates of decay of naturally occurring radioactive isotopes present in a rock or a fossil. Table 10.1 shows the half-life of six elements and the products of their decay. Figure 10.3 shows a general graph of the half-life of an element. Use the information in Table 10.1 and Figure 10.3 to answer Questions 1–3.

Table 10.1 Product of decay and half-life of elements used in radiometric dating.

Element	Product of decay	Half-life (years)
Carbon-14 (^{14}C)	Nitrogen-14 (^{14}N)	5730
Uranium-235 (^{235}U)	Protactinium-231 (^{231}Pa)	704 million
Uranium-234 (^{234}U)	Thorium-230 (^{230}Th)	246000
Potassium-40 (^{40}K)	Argon-40 (^{40}Ar)	1.25 billion
Thorium-232 (^{232}Th)	Lead-208 (^{208}Pb)	14 billion
Rubidium-87 (^{87}Ru)	Strontium-87 (^{87}Sr)	48 billion

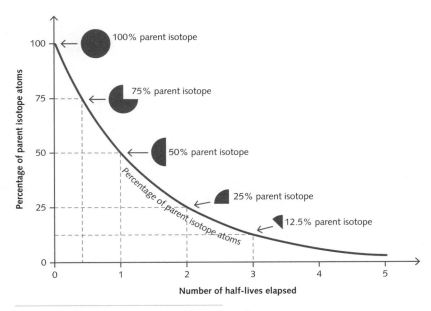

Figure 10.3 A graph of the half-life of an element

1 Which element would be best to use to radiometrically date metamorphic rocks of the following ages?

   a 10 000 years

   b 35 000 years

   c 200 000 years

   d 6.7 billion years

2 What does the term half-life mean?

3 In a 12 kg piece of igneous rock, 75% of the uranium-235 had decayed to protactinium-231. What is the approximate age of that rock?

4 Use the information in Figure 10.4 to estimate the age of the fossils shown in the rock layer. Provide your reasoning.

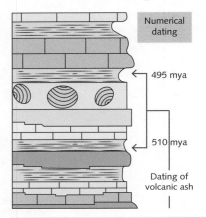

Figure 10.4

5 a Use the information in Figure 10.5 to rank layers A–E from youngest to oldest. Provide your reasoning.

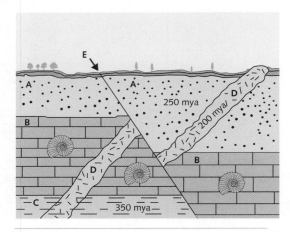

Figure 10.5

   b How old are the rocks from layer B? Provide your reasoning.

6 Estimate the age of each of the fossils shown in Figure 10.6. Provide your reasoning.

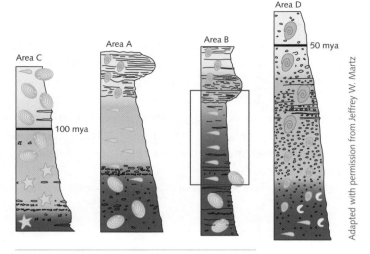

Figure 10.6

_____

_____

_____

_____

_____

10.2 Patterns in evolution

Key knowledge
Changes in species over time
- changes in species over geological time as evidenced from the fossil record: faunal (fossil) succession, index and transitional fossils, relative and absolute dating of fossils

10.2.1 Transitional fossils

Key science skills
Construct evidence-based arguments and draw conclusions
- evaluate data to determine the degree to which the evidence supports or refutes the initial prediction or hypothesis

Develop · PAGE 381

Read the following extract and use information from the extract and Figure 10.7, to answer the questions below.

Evolutionary pathways

Tetrapods are animals that have four limbs. Tetrapods include amphibians, reptiles and mammals, including humans. Tetrapods are thought to have evolved from lobe-finned fish (see Figure 10.7a). Over the years, fossils of lobe-finned fish and tetrapods have been found that have characteristics of both. This has shed light on some of the anatomical changes that these animals underwent, such as those that enabled breathing (lungs), hearing (ears) and feeding (teeth). What was lacking in the fossil record to complete this story was a complete pectoral fin fossil. This would make it possible to determine the evolution of one of the most important features for the transition from sea to land – hands.

(continued)

Elpistostege watsoni inhabited Earth during the Devonian Period between 416 million and 358 million years ago. Recently, a 1.57-metre fossil specimen of *E. watsoni* was found in Quebec, Canada. This animal was the missing link between lobe-finned fish and tetrapods. This crocodile-like creature had a flat head, long snout and small round eyes. It lived in shallow estuary waters. The most fascinating part of the find was the well-preserved pectoral (front) fins. They revealed the presence of phalanges that were organised into digits or fingers. This was the first evidence of finger-like bones that eventually evolved into hands (see Figure 10.7b). This discovery disproved the hypothesis that hands evolved in later land dwellers. It now seems that the fingers of vertebrates, including those on human hands, evolved from bony fin rays that support the fin into rows of digit bones in the fins of *Elpistostege* fishes. It is thought that as the fish began inhabiting shallow water, they relied more on the bones in their fins to support their weight. More small bones in the fin allows for more flexibility and spreading of weight.

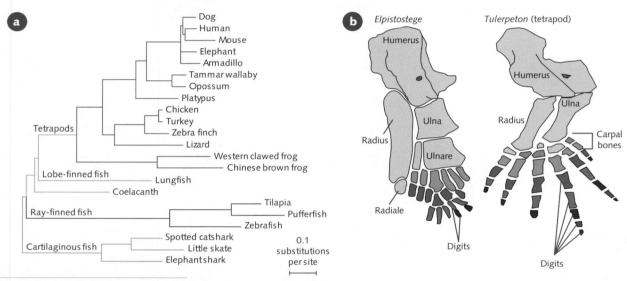

Figure 10.7 a A phylogenetic tree showing evolutionary pathway from fish to modern day animals; **b** fin bones of *Elpistostege* and *Tulerpeton* – a tetrapod

1   What is meant by a transitional fossil?

2   On Figure 10.7a, draw a circle to show where *Elpistostege watsoni* would have fitted into the phylogenetic tree.

3   What was the selection pressure that acted on the evolution of the pectoral fin bones in *Elpistostege watsoni*?

4   Considering the selection pressure identified in Question 3, which organisms would have been selected for?

5   Before the discovery of the pectoral fin fossil of *Elpistostege watsoni*, it was hypothesised that the vertebrate hand evolved on land. What is the status of that hypothesis in light of this new discovery?

6 Compare the tetrapod limb in Figure 10.7b to the human limb in Figure 10.18 on page 219.

    a State two similarities between them.

    b State two differences between them.

10.2.2 Divergent evolution

Divergent evolution is when differences between groups of organisms accumulate to a critical point that leads to speciation, the development of a new species. The cladogram in Figure 10.8 shows divergent evolution from a common ancestor. Use the information in Figure 10.8 to answer the questions below.

Consolidation of knowledge

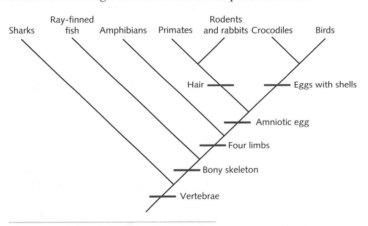

Figure 10.8 A cladogram showing divergent evolution from a common ancestor

1 Add the words 'common ancestor' to Figure 10.8 to show where the common ancestor for all the animals shown would be located on the cladogram.

2 Circle the point on the cladogram where the common ancestor for primates, rodents and rabbits would be located.

3 Which characteristic do all the animals shown on the cladogram possess?

4 List the characteristic(s) that both sharks and ray-finned fish possess.

5 List the characteristic(s) that both primates and rabbits possess.

6 List the characteristic that only crocodiles and birds possess.

7 Which two animals are more closely related: amphibians and primates or crocodiles and birds?

8 Which two animals are most distantly related?

10.2.3 Convergent evolution

PAGE 384 Convergent evolution occurs when unrelated organisms evolve similar adaptations in response to similar selection pressures. Figures 10.9 and 10.10 show examples of convergent evolution in pairs of animals and their associated phylogenetic trees.

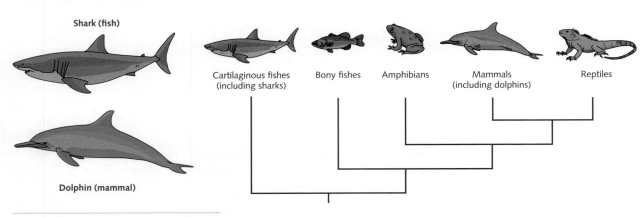

Figure 10.9 Convergent evolution in sharks and dolphins

1 Consider the shark and dolphin in Figure 10.9.

   a Identify the adaptation that has evolved in response to a similar selection pressure.

   b Identify the similar selection pressure acting on both animals.

   c Describe the evolutionary relationship between the two animals as shown by their phylogenetic tree.

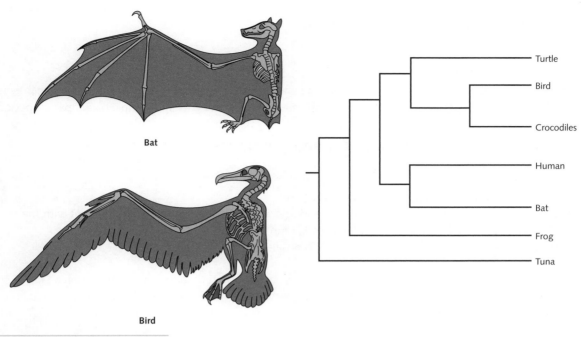

Figure 10.10 Convergent evolution in bats and birds

2   Consider the bat and bird in Figure 10.10.

   a   Identify the adaptation that has evolved in response to a similar selection pressure.

   b   Identify the similar selection pressure acting on both animals.

   c   Describe the evolutionary relationship between the two animals as shown by their phylogenetic tree.

10.3 Emergence of new species

Key knowledge

Changes in species over time
- evidence of speciation as a consequence of isolation and genetic divergence, including Galapagos finches as an example of allopatric speciation and *Howea* palms on Lord Howe Island as an example of sympatric speciation

10.3.1 Allopatric speciation and Galapagos finches

Key science skills

Construct evidence-based arguments and draw conclusions
- use reasoning to construct scientific arguments, and to draw and justify conclusions consistent with the evidence and relevant to the question under investigation
- discuss the implications of research findings and proposals

Peter and Rosemary Grant from Princeton University (USA) spent 40 summers on Daphne Major, an island in the Galapagos Islands, studying evolution firsthand. More than 100 years before the Grants, Charles Darwin visited the Galapagos Islands and documented the variety of finches on the different islands and the foods that they ate (Figure 10.11).

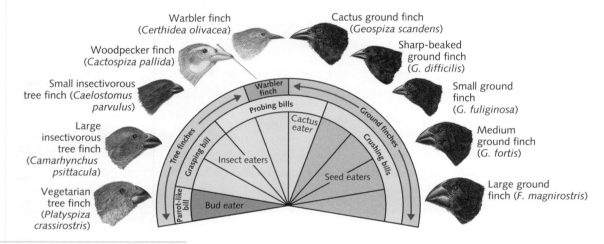

Figure 10.11 Finches observed by Charles Darwin on the Galapagos Islands

When the Grants began their investigation in 1973, the food available to the finches on Daphne Major was abundant soft small seeds. The Grants caught, measured and tagged the finches that lived on Daphne Major over a 40-year period. One of the measurements that they took was beak depth, using the method shown in Figure 10.12. The results obtained by the Grants for the year 1976 are shown in Figure 10.13.

Figure 10.12 The method used to measure beak depth

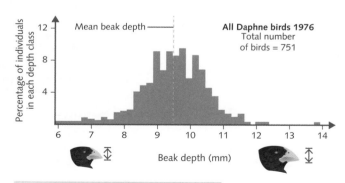

Figure 10.13 Beak depth data from Daphne Major in 1976

1 Using the information in Figure 10.13, describe the abundance of birds with different beak depths living on Daphne Major in 1976.

2 What relationship can you infer from Figure 10.13 in terms of the type of food available to the birds in 1976? Justify your answer by referring to Figure 10.11 to look for a relationship between beak depth and diet in finches.

3 Figure 10.14 shows temperature and rainfall data for Daphne Major from 1973 to 1978. Analyse the data and describe the relationships evident in the data.

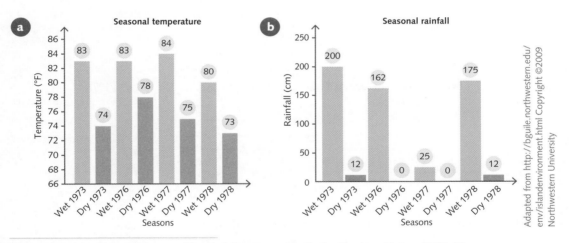

Figure 10.14 a Temperature and b rainfall measurements for Daphne Major, 1973–78

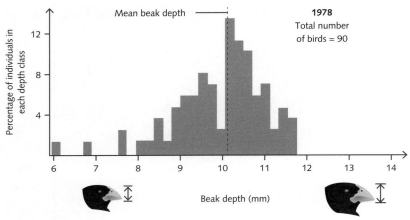

Figure 10.15 Beak depth data from Daphne Major in 1978

4   In 1978, the Grants recaptured any surviving tagged birds and remeasured their beak depth. This information is shown in Figure 10.15. Describe the abundance of birds with different beak depths living on Daphne Major in 1978.

5   Compare the beak depth data from 1976 (Figure 10.13) to that of 1978 (Figure 10.15). Describe any changes.

6   What can you infer from this data in terms of the type of food available to the birds in 1978?

7   Using all the data presented, explain how this change to beak depth might have come about.

8   What might this change to beak depth mean for future generations of finches on Daphne Major?

9   Use the information that you have learnt about the finches on Daphne Major to explain the allopatric speciation of finches across the Galapagos Islands (Refer to Figure 10.11.)

_____

_____

_____

_____

_____

_____

10.3.2 Sympatric speciation and *Howea* palms

Key science skills
Construct evidence-based arguments and draw conclusions
*   use reasoning to construct scientific arguments, and to draw and justify conclusions consistent with the evidence and relevant to the question under investigation

Lord Howe Island is a volcanic island 580 km off the coast of south-east Queensland. Figure 10.16 shows the flowering times of the two *Howea* species that live on Lord Howe Island. *Howeas* are flowering plants in the palm family. Green represents 198 curly palms (*Howea belmoreana*) and grey represents 177 kentia palms (*Howea forsteriana*).

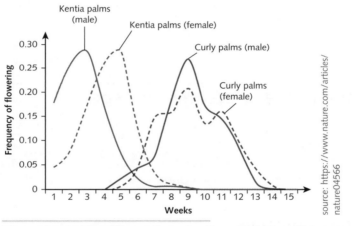

Figure 10.16 The different flowering times of the two *Howea* species

Figure 10.17 shows the soil and altitude preferences of the two species of *Howea* palms.

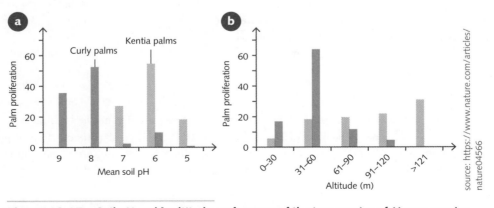

Figure 10.17 a Soil pH and b altitude preferences of the two species of *Howea* species

Using the evidence provided in Figures 10.16 and 10.17 and your knowledge of the process of natural selection, construct a reasoned argument supported by evidence that answers the following question.

Is the speciation of curly and kentia Howea palms on Lord Howe Island an example of sympatric speciation?

10.4 Determining the relatedness of species

Key knowledge

Determining the relatedness of species
- evidence of relatedness between species: structural morphology – homologous and vestigial structures; and molecular homology – DNA and amino acid sequences

10.4.1 Homologous structures

Key science skills

Generate, collate and record data
- systematically generate and record primary data, and collate secondary data, appropriate to the investigation, including use of databases and reputable online data sources

Construct evidence-based arguments and draw conclusions
- use reasoning to construct scientific arguments, and to draw and justify conclusions consistent with the evidence and relevant to the question under investigation

Develop

Homologous structures are similar physiological structures in different organisms that can be explained by the organisms' descent from a common evolutionary ancestor. Figure 10.18 shows a generalised pentadactyl limb surrounded by the pentadactyl limbs of several animals. The wing of a bird, the wing of a bat, the leg of a crocodile, the flipper of a whale, the forelimbs of a cat and a lizard, and the arm of a human all have different functions and appear superficially different. However, they all have the same basic skeletal structure: a pentadactyl limb with a hand or foot with five fingers or toes.

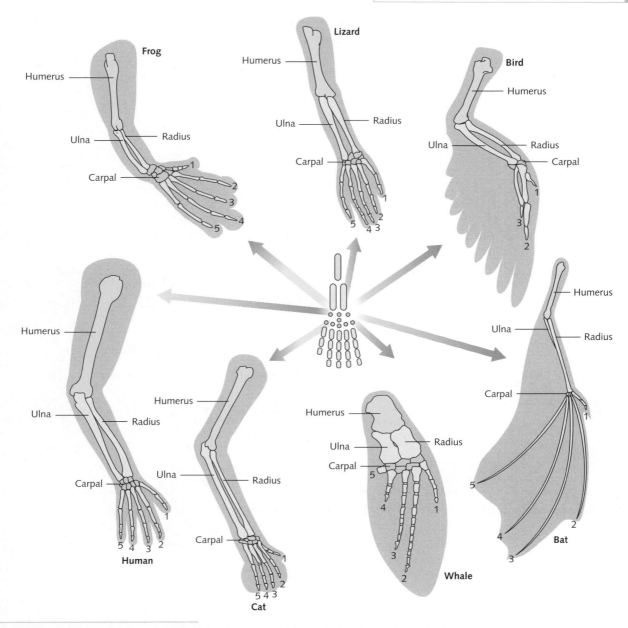

Figure 10.18 Generalised pentadactyl limb (centre) and pentadactyl limbs of different animals

1   Using Figure 10.18, complete Table 10.2 by writing in the number of bones that form each of the listed structures.

Table 10.2 Number of bones in pentadactyl limb

Structure	Generalised pentadactyl limb	Animal						
		Frog	Lizard	Bird	Bat	Whale	Cat	Human
Upper arm								
Forearm								
Wrist								
Hand								
Digit 1								
Digit 2								
Digit 3								
Digit 4								
Digit 5								

2 a Refer to your data in Table 10.2. How do the number of bones in each structure for the animals listed compare with the number in the generalised pentadactyl limb?

b Other than numbers of bones, what other differences did you find in the limb structures? Explain.

c Can you suggest any reasons for the differences in the limb structure for each animal? Explain.

d What advantage might these differences offer to the species concerned?

3 Although there are some differences in the structures of these animals, there are still a lot of similarities. Suggest an explanation as to why these similarities can be found in so many different species even though they may occupy different habitats.

4 What conclusion can you draw from the data provided?

10.5   Molecular evidence for relatedness between species

Key knowledge
Determining the relatedness of species
- evidence of relatedness between species: structural morphology – homologous and vestigial structures; and molecular homology – DNA and amino acid sequences
- the use and interpretation of phylogenetic trees as evidence for the relatedness between species

10.5.1 The concept of molecular homology

Key science skills
Construct evidence-based arguments and draw conclusions
- use reasoning to construct scientific arguments, and to draw and justify conclusions consistent with the evidence and relevant to the question under investigation

Develop
PAGE 399

Cytochrome c is an essential protein in the electron transport chain of cellular respiration. Cytochrome c has a primary structure of 104 amino acids. Table 10.3 shows a section of cytochrome c for eight species, from humans to yeast. Use the molecular evidence provided in Table 10.3 to analyse these sequences of amino acids shown and then answer the questions below.

Table 10.3 Molecular homology of cytochrome c

Organism	Amino acid position																					
	1	2	3	4	5	6	7	8	9	10	11	12	13	14	15	16	17	18	19	20	21	22
Human	Gly	Asp	Val	Glu	Lys	Gly	Lys	Lys	Ile	Phe	Ile	Met	Lys	Cys	Ser	Gln	Cys	His	Thr	Val	Glu	Lys
Chimpanzee	Gly	Asp	Val	Glu	Lys	Gly	Lys	Lys	Ile	Phe	Ile	Met	Lys	Cys	Ser	Gln	Cys	His	Thr	Val	Glu	Lys
Pig	Gly	Asp	Val	Glu	Lys	Gly	Lys	Lys	Ile	Phe	Val	Gln	Lys	Cys	Ala	Gln	Cys	His	Thr	Val	Glu	Lys
Chicken	Gly	Asp	Ile	Glu	Lys	Gly	Lys	Lys	Ile	Phe	Val	Gln	Lys	Cys	Ser	Gln	Cys	His	Thr	Val	Glu	Lys
Dogfish	Gly	Asp	Val	Glu	Lys	Gly	Lys	Lys	Val	Phe	Val	Gln	Lys	Cys	Ala	Gln	Cys	His	Thr	Val	Glu	Asn
Drosophila	Gly	Asp	Val	Glu	Lys	Gly	Lys	Lys	Leu		Val	Gln	Arg		Ala	Gln	Cys	His	Thr	Val	Glu	Ala
Wheat	Gly	Asp	Pro	Asp	Ala	Gly	Ala	Lys	Ile	Phe	Lys	Thr	Lys	Cys	Ala	Gln	Cys	His	Thr	Val	Asp	Ala
Yeast	Gly	Asp	Ala	Lys	Lys	Gly	Ala	Thr	Leu	Phe	Lys	Thr	Arg	Cys	Glu	Leu	Cys	His	Thr	Val	Glu	Lys

1 The more similarity there is between the cytochrome c from different species, the more recently the species have evolved from a common ancestor. Determine the order of evolution from a common ancestor for the eight animals shown in Table 10.3, starting with most distant and ending with the most recent.

2 Which amino acid positions are probably crucial, and which is probably the least crucial? Explain your reasoning.

3 Do you think this evidence supports other knowledge you have of relationships between these species? Explain your reasoning.

10.5.2 Assembling phylogenetic trees

PAGE 404

Key science skills
Generate, collate and record data
- organise and present data in useful and meaningful ways, including schematic diagrams, flow charts, tables, bar charts and line graphs

Reinforce

Phylogenetic trees include both cladograms and phylograms. Figure 10.19 is a cladogram showing the evolution of 10 animal groups from a common archosaur ancestor.

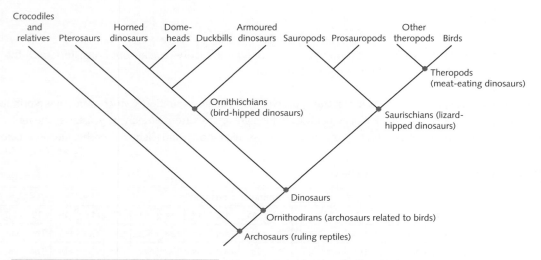

Figure 10.19 A cladogram showing evolution of a common archosaur ancestor

1   Use the information in Figure 10.19 to construct a horizontal cladogram in the space below.

2 **a** How many clades are represented in the cladogram?

_____

b Are the following pairs monophyletic?

   i Sauropods and prosauropods

_____

   ii Prosauropods and other theropods

_____

   iii Horned dinosaurs and Dome-heads

_____

   iv Crocodiles and pterosaurs

_____

   v Dome-heads and armoured dinosaurs

_____

c State the name of the common ancestor for each of the following pairs.

   i Crocodiles and pterosaurs

_____

   ii Dome-heads and armoured dinosaurs

_____

   iii Sauropods and birds

_____

   iv Duckbills and birds

_____

10.6 Chapter review

10.6.1 Key terms

PAGE 415

1 Define the following terms.

a Radioactive decay

b Absolute dating

c Reproductive isolation

d Speciation

e Allopatric speciation

f Sympatric speciation

g Divergent evolution

h Adaptive radiation

i Convergent evolution

j Vestigial structure

k   Homologous structure

_____

_____

l   Analogous structure

_____

_____

m   Molecular homology

_____

_____

n   Phylogenetic tree

_____

_____

o   Cladogram

_____

_____

2   Write the definitions of five other key terms of your choice from this topic.

_____

_____

_____

_____

_____

_____

_____

_____

_____

_____

_____

_____

_____

_____

_____

_____

Chapter review (continued)

10.6.2 Exam practice

PAGE 417

Exam practice

The following information relates to Questions 1 and 2.

The following phylogenetic tree summarises the evolutionary relationships between fish species.

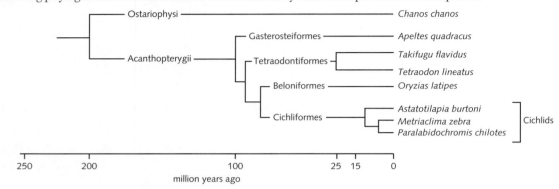

1 ©VCAA 2019 Q24 (adapted) MEDIUM

Apeltes quadracus is most closely related to

A   *Chanos chanos.*

B   *Tetraodon lineatus.*

C   *Paralabidochromis chilotes.*

D   *Astatotilapia burtoni.*

2 ©VCAA 2019 Q25 (adapted) EASY

Which one of the following statements is correct?

A   Tetraodontiformes diverged to form two distinct species 25 million years ago.

B   *Chanos chanos* was the last species to diverge from the most distant common ancestor.

C   Gasterosteiformes, beloniformes and cichliformes do not share a common ancestor.

D   *Metriaclima zebra* and *Astatotilapia burtoni* are the two most closely related species on this phylogenetic tree.

The following information relates to Questions 3 and 4.

Cytochrome c is a protein that consists of 104 amino acids. Many of the 104 amino acid sites are exactly the same across a large range of organisms. However, there are some amino acid differences at certain sites. It is hypothesised that different organisms, all containing cytochrome c proteins, descended from a primitive microbe that lived more than 2 billion years ago. Table 10.4 uses the three-letter codes for the amino acids found at five sites for each organism.

Table 10.4 Molecular homology of cytochrome c

Organism	Site 1	Site 4	Site 11	Site 15	Site 22
Human	Gly	Glu	Ile	Ser	Lys
Pig	Gly	Glu	Val	Ala	Lys
Dogfish	Gly	Glu	Val	Ala	Asn
Chicken	Gly	Glu	Val	Ser	Lys
Drosophila	Gly	Glu	Val	Ala	Ala
Yeast	Gly	Lys	Val	Glu	Lys
Wheat	Gly	Asp	Lys	Ala	Ala

3 ©VCAA 2016 Q39 (adapted) EASY

Using only the data for the molecular homology of cytochrome c shown in Table 10.4, determine which one of the following organisms is most closely related to the pig.

A Wheat
B Chicken
C Human
D Yeast

4 ©VCAA 2016 Q40 (adapted) MEDIUM

Using only the data for the molecular homology of cytochrome c shown in Table 10.4, determine which pair of organisms is most distantly related to wheat.

A Dogfish and *Drosophila*
B *Drosophila* and yeast
C *Drosophila* and pig
D Human and yeast

5 ©VCAA 2011 E2 Q23 (adapted) MEDIUM

Two possible phylogenetic relationships between eight groups of flowering plants are shown in the following diagrams.

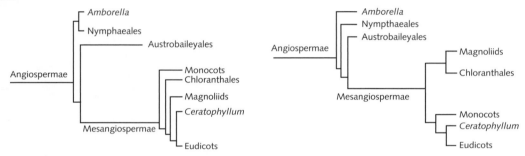

One similarity between the two diagrams is that

A monocots diverged before magnoliids
B *Ceratophyllum* and eudicots are the latest group to diverge
C eudicots were the first group to diverge from Mesangiospermae
D *Ceratophyllum* and eudicots diverged from monocots.

6 ©VCAA 2019 Section B Q7 The following table provides information on three species. (7 marks)

Species	Goldfish	Dolphin	Penguin
	iStock/Getty Images Plus/thumb	Shutterstock.com/ chonlasub woravichan	Shutterstock.com/Eric Isselee
Size	Tiny freshwater fish (10–50 g)	Medium-sized aquatic mammal (150–600 kg)	Small marine bird (2–30 kg)

a These three species are not closely related but have common features that have evolved in response to a common selection pressure.
Identify the type of evolution involved. Justify your response. 2 marks

_____

_____

_____

b   In 2015, a 5 cm fossil of a neck bone from a delicately built dinosaur was uncovered at Cape Otway. It is thought to belong to a dinosaur known as an *Elaphrosaur*, which would have grown to around 2 m in length. The site where the fossil was found would have been a river when the *Elaphrosaur* was alive.

i   There are no living specimens of *Elaphrosaur*, so it is known as an extinct species.
Give **one** possible cause of its extinction.   1 mark

_____

_____

ii   Although *Elaphrosaur* is extinct, palaeontologists are piecing together its skeleton from fossil fragments. Describe what would have occurred to lead to the fossilisation of the *Elaphrosaur*, from when it died to when the fossils were discovered.   4 marks

_____

_____

_____

_____

_____

_____

_____

_____

Human change over time

11

Remember

PAGE 424

In Unit 4 of VCE Biology, you learnt about evolution by natural selection and how this can lead to speciation. In this final chapter, you will finish with a fascinating area of biology that will require you to draw on a broad and varied knowledge and skill base. Use the questions below to focus your attention on some core ideas. Ensure you have a solid grasp of these key ideas by taking notes and identifying areas that require further reading.

1   Differentiate between allopatric and sympatric speciation.

2   Does the fossil record present a complete record of life in the past?

3   What is a fossil and what can it tell us about the past?

4   Using carbon-14 as an example, explain how radioisotope dating works.

5   List the features that all mammals possess.

6   Which two animals are most closely related: horse and cow, horse and bird, or bird and fish? Explain your choice.

7   What does a phylogenetic tree depict? What does a node in a phylogenetic tree represent?

_____

_____

8   What can a phylogram tell you that a cladogram cannot?

_____

_____

_____

11.1   Taxonomy of modern humans

Key knowledge
Human change over time
- the shared characteristics that define mammals, primates, hominoids and hominins

11.1.1 Taxonomy

PAGE 425

Key science skills
Generate, collate and record data
- organise and present data in useful and meaningful ways, including schematic diagrams, flow charts, tables, bar charts and line graphs

Construct evidence-based arguments and draw conclusions
- use reasoning to construct scientific arguments, and to draw and justify conclusions consistent with the evidence and relevant to the question under investigation

Develop

Like all other animals, humans can be classified using a taxonomic scheme. This activity requires you to interpret diagrams to find the answers to questions. This is an important skill and one that you need to practise.

1   Using information from your VCE Biology textbook and Figure 11.1, complete Table 11.1.

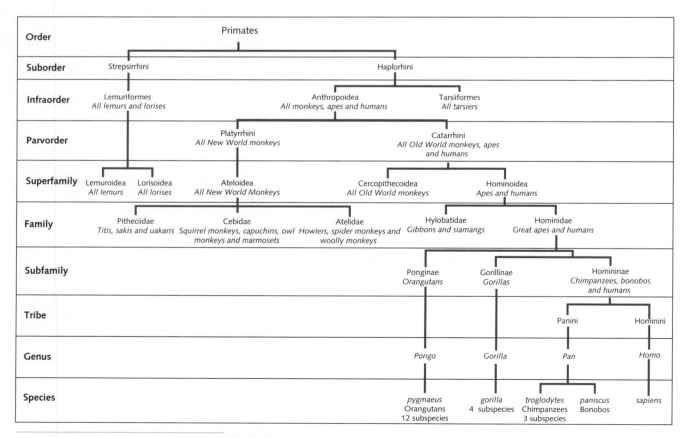

Figure 11.1 Taxonomy of primates superimposed on a cladogram

Table 11.1 Taxonomy of modern humans

Level of taxonomy	Humans	Features that all members of this level of taxonomy share	Which of the following animals also belong to this taxonomic level? • lemur • mandrill • chimpanzee • ape • orangutan
Kingdom			
Phylum			
Class			
Order			
Family			
Genus			
Species			

2   Figure 11.1 shows the position of modern humans on a cladogram of all primates with taxonomy information superimposed. Additional levels of classification are introduced to the taxonomy classifications (e.g. superfamily, tribe) to better represent the evolutionary relationships.

a   At which taxonomic level does the human line branch from the lemurs and lorises?

b   At which taxonomic level does the human line branch from the New World monkeys?

c At which taxonomic level does the human line branch into a line that contains only *Homo sapiens*?

d List the families that make up the order Primates.

e Are humans more closely related to orangutans, gorillas or chimpanzees?

f Are orangutans and gorillas more or less closely related than chimpanzees and bonobos? Explain your reasoning.

g What features do all members of the family Hominidae possess?

h List the features that set *Homo sapiens* apart from *Pan troglogytes*.

3 Figure 11.2 shows a phylogenetic tree with molecular clock estimates for divergence between major hominoid lineages.

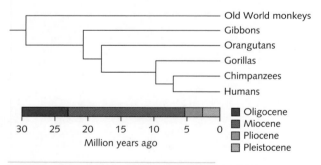

Figure 11.2 Phylogenetic tree for major hominoid lineages

a When did the Old World monkey line diverge from the ape and human line?

b When did the human line diverge from its nearest relative?

11.2   Adaptations that define humans

11.2.1 Adaptations for bipedalism

Most hominoids are quadrupedal, which means they use all four limbs for locomotion. Humans are bipedal, which means they walk on their hind limbs. Bipedalism has only been possible through evolutionary changes to the skeleton.

Consolidation of knowledge

TB PAGE 433

Annotate Figure 11.3 to indicate the features of the human skeleton that enable bipedal gait.

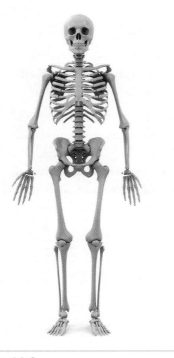

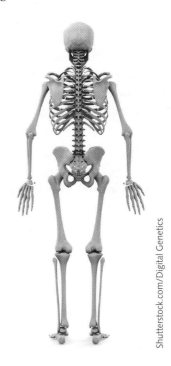

Shutterstock.com/Digital Genetics

Figure 11.3

11.2.2 Evolution of two-legged walking

Develop

TB PAGE 435

Read the following article about the emergence of bipedalism and answer the questions below.

Did ancient apes walk on two feet?

Fossilised remains of four ancient apes have been discovered in a Bavarian clay pit, and it is destabilising the scientific world. The ape, *Danuvius guggenmosi*, lived around the area of what is now known as Germany about 11.6 million years ago. Its skeletal remains show strong arms, hands and feet, which suggest that it used to grip branches and swing through trees, but the leg bones and vertebrae suggest it was also adapted to moving around on two feet. The shapes of the vertebrae suggest that the lower back was long and flexible, and the knees and ankles indicate weight-bearing adaptations. These are all features seen in humans, who stay balanced by pulling the weight of their torso over their hips to remain upright. In the scientific world, bipedalism, or walking on two feet, is thought to be a characteristic unique to hominins.

Add to this the discovery of another ancient European ape, *Rudapithecus hungaricus*, which showed pelvic adaptations to an upright stance and had a long and flexible lower back. Like *D. guggenmosi*, it could have been a tree-dwelling biped.

Ardipithecus ramidus, a 4.4-million-year-old hominin, was clearly bipedal and there are suggestions of bipedal gait in species as old as 7 million years. Seven million years ago is important in evolutionary terms because this is when it is thought that the evolutionary tree branch that contains modern chimpanzees and bonobos split from the hominin branch. However, the appearance of the skeleton of *D. guggenmosi* suggests that bipedalism existed before this split occurred. If the scientists are correct about *D. guggenmosi*, then bipedalism evolved many millions of years before humans even appeared. Do scientists have to redefine what it means to be a hominin?

Some scientists argue against such a conclusion, saying that the available remains of the fossilised spine are fragmentary and deformed.

So, what conclusions can be drawn? Is there a direct line of descent from *D. guggenmosi* to modern humans, or did today's knuckle-walking chimps and gorillas evolve from a bipedal ancestor? The big question is whether bipedalism evolved more than once, and what conditions were around to select for bipedal gait.

1   Is the material presented in this article opinion, anecdote or evidence? Provide your reasons.

_____

_____

_____

2   What is the existing hypothesis that has been refuted by this new evidence?

_____

_____

3   What is the new hypothesis that is presented in this article?

_____

_____

4   What evidence has been put forward to support the new hypothesis?

_____

_____

_____

_____

5 Add a star shape to Figure 11.2 on page 232 to show where *D. gugenmosi* would be located on the phylogenetic tree.

6 a Explain why some scientists are reluctant to support the new hypothesis.

_____

_____

_____

b What types of evidence would need to be provided for these scientists to change their minds?

_____

_____

_____

7 What would be the implications of this finding if the new hypothesis were found to be supported?

_____

_____

_____

_____

_____

_____

11.2.3 Human ingenuity

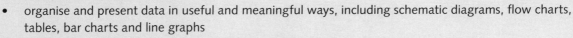

Key science skills

Generate, collate and record data
- organise and present data in useful and meaningful ways, including schematic diagrams, flow charts, tables, bar charts and line graphs

Construct evidence-based arguments and draw conclusions
- use reasoning to construct scientific arguments, and to draw and justify conclusions consistent with the evidence and relevant to the question under investigation

TB PAGE 439

Develop

Human ingenuity is the ability of the human mind to think at a high level to rationalise and find solutions to problems. Ingenuity influences how we work, interact and communicate with one another, use our environment, and transform our environment and those of other species. Ingenuity also gives us the ability to plan and think about the consequences of our actions. Human ingenuity would not be possible had it not been for two evolutionary changes to our skeleton: our hands and our skull.

Human hands

The one feature that set humans apart from the other primates was the *bipedal stride*, which freed the hands and gave humans the potential to carry weapons and make tools. The grasping hand was a major advance and enabled humans to undertake precision work directly in front of their eyes.

Figure 11.4 The precision grip enabled humans to produce tools

A characteristic of all primates is the opposable thumb, allowing a precision grip. An opposable thumb can touch the tips of all other fingers of that hand. To demonstrate the importance of an opposable thumb, try the exercise below.

1 Use a pen or pencil to write the following sentence in the space provided.

The opposable thumb allows me to hold a pen so I can write.

2 Use sticky tape to hold your thumb to your hand so that you cannot move your thumb. Use a pen or pencil to write the following sentence in the space provided.

The opposable thumb allows me to hold a pen so I can write.

3 What did you discover about your ability to perform a precision task such as writing with and without your opposable thumb?

Expansion of the cranium

The precision grip with the opposable thumb, coupled with a large brain compared with other primates gave humans advanced cognitive capacity. Table 11.2 shows the brain size of eight primates.

Table 11.2 Comparison of primate brain sizes

Primate	Average brain volume (cm³)	Evolutionary distance in relationship to humans (scale 1–8)
Prosimian	12.6	
Lemur	24.0	
New World monkey	34.1	
Old World monkey	89.1	
Lesser ape	97.5	
Great ape	316.7	
Chimpanzee	393.0	
Human	1350.0	1

4 Graph the data in the first two columns of Table 11.2 using the axes below. Make sure that you meet all the criteria for drawing a scientific graph.

5 Refer to Figure 11.1 on page 230 and use the scale 1–8 to complete the last column in Table 11.2.

6 Comment on the trend shown in the graph and relate this to the evolutionary distance of each primate from *Homo sapiens*.

_____

_____

_____

7 What capacities and abilities does the large brain size allow in humans?

_____

_____

_____

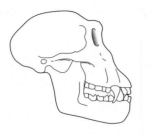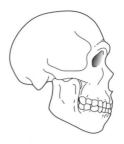

Figure 11.5 A comparison between skulls of humans and skulls of chimpanzees

8 Using the information in Table 11.2 and Figure 11.5, state the differences in skull shape between chimpanzees and humans and account for each difference.

_____

_____

_____

_____

_____

_____

9 Using the data, evidence and information provided in this activity, draw a conclusion about the defining features of humans compared with other primates.

_____

_____

_____

_____

_____

_____

_____

_____

11.3   Meet the ancestors

Key knowledge
Human change over time
* evidence for major trends in hominin evolution from the genus _Australopithecus_ to the genus _Homo:_ changes in brain size and limb structure
* the human fossil record as an example of a classification scheme that is open to differing interpretations that are contested, refined or replaced when challenged by new evidence, including evidence for interbreeding between _Homo sapiens_ and _Homo neanderthalensis_ and evidence of new putative _Homo_ species

11.3.1 From *Australopithecines* to *Homo sapiens*

Table 11.3 shows the average brain volume of different species in the evolution of *Homo sapiens*.

Table 11.3 Brain volume of some species in the evolution of *Homo sapiens*

Species	Average brain volume (cm³)
Australopithecus afarensis	385
Australopithecus africanus	415
Homo habilis	600
Homo erectus	900
Early *Homo sapiens*	1150
Modern *Homo sapiens*	1350

1 Graph the data in Table 11.3 using the axes below. Make sure that you meet all the criteria for drawing a scientific graph.

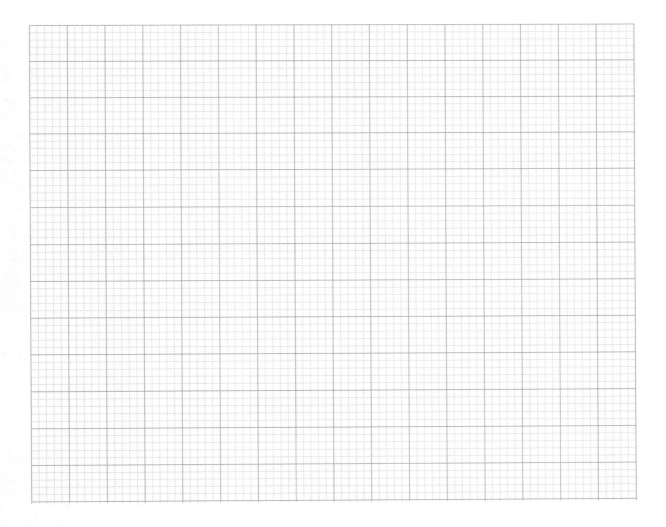

2 On your graph, place a cross through the species that is not directly related to *Homo sapiens*.

3 On your graph, draw a circle around the species that is thought to be transitional between *Australopithecus* and *Homo*.

Since Charles Darwin first proposed that humans had evolved from ape-like ancestors, there has been great interest in the ancestry of the human race. But how much information can we obtain from the fossil record, and how sure are we of our interpretation of the information?

In 2003, a group of Australian and Indonesian scientists found the remains of an early human on the island of Flores in Indonesia. This early human has been given the name *Homo floresiensis* and was about 1 metre tall. *Homo floresiensis* looked nothing like the humans of today; it had a diminutive stature, small brain (about 400 cm³), big teeth, receding chin and a pelvis that flared out to the side.

Scientists are debating whether the remains are a new species, the same species as modern humans or a subspecies of *Homo sapiens*. Some scientists have speculated that the small brain size means that *Homo floresiensis* descended from an ancestor somewhere between the early hominid *Australopithecus* and the genus *Homo*.

4 a Add *Homo floresiensis* to your graph in Question 1. Where is it located?

_____

b What are some of the key body structures found in the remains of *Homo floresiensis* that you would want to examine to determine whether it is a close relative of *Homo sapiens* or a more ape-like relative? Explain.

_____

_____

_____

_____

_____

c How could a distinct population of *Homo floresiensis* evolve when other human-like populations also existed around that time?

_____

_____

_____

_____

d Dating of the remains suggested that *Homo floresiensis* may have existed at the same time as other more modern humans. Taking this into account, can you suggest reasons why *Homo floresiensis* became extinct?

_____

_____

_____

_____

_____

_____

e If the *Homo floresiensis* remains came from only one individual, why do you think scientists would be considering that the remains may represent the same species as modern humans?

_____

_____

_____

_____

f What further evidence, other than brain size, would be needed to form a clearer understanding of where *Homo floresiensis* fits into the evolutionary tree of humans? Explain your answer.

_____

_____

_____

_____

_____

_____

11.4   Modern humans and Neanderthals

Key knowledge

Human change over time

- the human fossil record as an example of a classification scheme that is open to differing interpretations that are contested, refined or replaced when challenged by new evidence, including evidence for interbreeding between *Homo sapiens* and *Homo neanderthalensis* and evidence of new putative *Homo* species
- ways of using fossil and DNA evidence (mtDNA and whole genomes) to demonstrate the migration of modern human populations around the world, including migration of Aboriginal and Torres Strait Islander populations and their connection to Country and Place.

11.4.1 Evolution of modern humans

Key science skills

Generate, collate and record data

- organise and present data in useful and meaningful ways, including schematic diagrams, flow charts, tables, bar charts and line graphs

TB
PAGE 454

Reinforce

Fossil evidence suggests that early humans evolved in central Africa. It is thought that drought forced them to migrate to the east coast of Africa and then northwards in what is known as the Out of Africa, or recent single origin, hypothesis.

1 On the map of the world in Figure 11.6 on page 242, trace the routes taken by our early ancestors as they left Africa to inhabit the rest of the world.

2 Provide evidence for each of your routes in the form of fossil, cultural and molecular evidence.

3 Identify locations where early humans would have interbred with other *Homo* species such as *Homo neanderthalensis*.

4 Where possible, provide the approximate dates when early humans arrived at different continents.

Figure 11.6

11.4.2 Australian settlement

Key science skills
Construct evidence-based arguments and draw conclusions
- use reasoning to construct scientific arguments, and to draw and justify conclusions consistent with the evidence and relevant to the question under investigation

Develop

TB
PAGE 458

Read the following extracts and use the information from the extracts and your textbook to answer the questions below.

Researchers at the University of Adelaide analysed the mitochondrial DNA (mtDNA) from 111 hair samples stored at the South Australian Museum. The results showed that a single founding population arrived in Australia via New Guinea 50 000 years ago. At this time, Australia and New Guinea were connected by a land bridge forming a landmass called Sahul. More than 8000 years ago, sea levels began to rise and cut off New Guinea from mainland Australia.

The results also showed that Aboriginal populations in Australia had almost no geographic movement for 50 000 years, enabling them to form sacred and cultural connections to their Country. Although the environment around them changed significantly, they were able to survive in the area by using a fixed set of resources. This has not been seen anywhere else in the world. Fossils such as stone tools, rock art, shell middens (where the remains of meals were put), charcoal deposits and human skeletal remains provide evidence of a complex culture. Rock art was used to tell stories to convey their knowledge of the land and beliefs and to pass this onto the next generation. The rock art also provides clues about the types of animals that were hunted, tools that were used for hunting and the clothing and body adornments that were worn. Some of the earliest rock art, dating to around 50 000–60 000 years ago, can be found in Kakadu in the Northern Territory and the Kimberleys in Western Australia.

Aboriginal DNA study reveals 50,000-year story of sacred ties to land, by Melissa Davey, 9 March 2017, *The Guardian*, Australia news. Copyright Guardian News & Media Ltd 2020

Aunty Zeta Thomson, Victorian Wurundjeri and Yorta Yorta Aboriginal Elder says that '… over thousands and thousands of years, we are more conscious than ever that races are very special and we need to be guardians of maintaining culture and what is there for us and for the next generation … my Country is part of my people … it's like a healing for a lot of people, if you go to a certain area, like where my grandfather was born, or spend a lot of time anywhere in the Country, but in particular along the Murray River, it is a very healing place for us. Aboriginal people follow a moiety, where they must marry someone from outside of their clan. Bunji is a creative spirit ancestor, who is the maker of all things, who comes in the form of a wedge-tailed eagle. And Waa who is the keeper of the wind and the waterways, and who comes in the form of a black crow.'

When Europeans settled in Australia, they forced the Aboriginal people off Country and removed them from their families, scattering them all over Australia and creating a disconnection between people and land.

Aunty Zeta says that '… they [Aborigines] were all stopped from talking their language because it was taboo. The people (Europeans) didn't know what they were talking about so they stopped them by whipping them by not giving them rations. They [Europeans] stopped any ceremonies because it was heathen to them. The ceremonies that had been passed on for thousands of years all pretty much just stopped. The way of life of Aboriginal people was stopped dead in its tracks after all their time living it, and so the Aboriginal people were trying to live in two worlds, the Aboriginal world and the non-Aboriginal world to survive.'

VICscience VCE BIOLOGY SKILLS WORKBOOK UNITS 3 & 4

Genes on mtDNA are referred to as a haplotype as they are a group of genes that are inherited together from a single parent. Individuals that share haplotypes are called haplogroups. The hair samples analysed by the University of Adelaide researchers identified four main haplogroups of mtDNA: P, O, S and M. Evidence of the distribution of these haplogroups in Australia is shown in Figure 11.7.

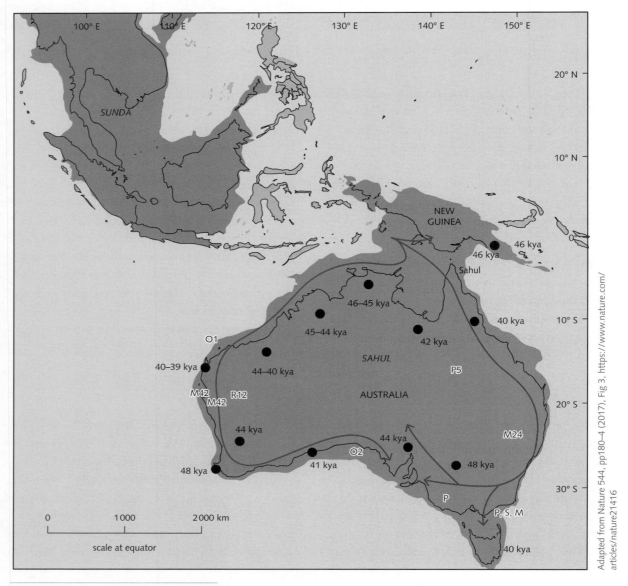

Adapted from Nature 544, pp180–4 (2017), Fig 3, https://www.nature.com/articles/nature21416

Figure 11.7 The distribution of mitochondrial DNA haplogroups in Australia

1   Explain the benefits of analysing mtDNA rather than nuclear DNA to investigate migration patterns.

2   Explain why the migration patterns have been shown to be predominantly down the east and west coasts of Australia.

3   Which haplogroup(s) moved down the west coast?

4 Which haplogroup(s) moved down the east coast?

5 What geographical feature prevented the east coast and west coast haplogroups from intermingling?

6 Which haplogroup would you expect to be most dominant in central Australia?

7 The mtDNA haplogroups S, P, Q, R, O, M, N and R can be found in Aboriginal people across Australia. Explain why these haplogroups are not all represented in the study by scientists at the University of Adelaide.

8 How do fossils demonstrate the Aboriginal and Torres Strait Islander connection to Country and Place?

9 How did European colonisation affect the relationship of Aboriginal and Torres Strait Islander peoples with their Country?

10 Comment on the quality of the different types of evidence presented in the extracts on page 243. Do you have any reasons to query what has been presented?

11.4.3 Relationship between modern humans and Neanderthals

Key science skills
Construct evidence-based arguments and draw conclusions
- identify, describe and explain the limitations of conclusions, including identification of further evidence required

Develop

The following extract describes the identification of bone fragments from an archaeological site in eastern Europe. Read this extract carefully and use the information to answer the questions below.

Bacho Kiro in Bulgaria is the site of an archaeological excavation that was first excavated in 1938 and then again in the 1970s. In the latter dig, archaeologists uncovered animal bones, stone and bone tools, beads and pendants and ancient human remains. Among the discoveries, which included more than 1200 bone and teeth fragments, only a single molar could be identified as being from a modern human. Analysis of the protein structure from other fragments revealed that six of the other items were remains from modern humans. The tools, beads and pendants found at Bacho Kiro closely resembled artefacts that had been found at a Neanderthal site in western Europe, which had been dated several thousand years later. The evidence from Bacho Kiro led archaeologists to conclude that modern humans had arrived in eastern Europe and interacted with Neanderthals for a full 5000 years longer than previously thought. The questions that remain unanswered are whether there had been an 8000-year overlap between modern humans and Neanderthals, and, if so, why the Neanderthals become extinct.

1 How does this new evidence alter our ideas about the overlap between Neanderthals and modern humans?

2 What evidence was found that indicated that modern humans were present at this Neanderthal site?

3 Provide one hypothesis to explain the extinction of Neanderthals.

4 What sort of evidence would you need to support your hypothesis?

5 Europeans and Asians have approximately 1–4% Neanderthal DNA in their genomes but sub-Saharan Africans have none. Explain why.

PAGE 468

11.5   Chapter review

11.5.1 Know your key terms

1   Use the space below to arrange these terms in order of chronological occurrence, starting with the youngest. A term may sit beside another to show that they existed at the same time. Add annotations to each term to show dates, areas and evidence used to understand their life and culture.

Australopithecus

Homo

Australopithecus afarensis

Homo erectus

Homo sapiens

Homo neanderthalensis

Danuvius guggenmosi

Homo habilis

Homo floresiensis

Denisovan

2   Distinguish between the following pairs of terms.

   a   Cognitive capacity and cranial capacity

   b   Haplogroup and haplotype

   c   Hominin and hominoid

   d   Precision grip and opposable thumb

   e   Bipedal and quadrupedal

   f   Saggital crest and saggital keel

   g   Oviparous and viviparous

Chapter review (continued)

PAGE 469

Exam practice

11.5.2 Exam practice

1 ©VCAA 2018 Q37 (adapted) Members of the order Primates are mammals. Which combination of features is common to all primates and distinguishes them from other mammals?

- **A** Forward-facing eyes, sloping forehead, feed their young with milk
- **B** Parabolic jaw, tail, four chambers in heart
- **C** Hands and feet with five digits, opposable thumbs, see in colour
- **D** Even-sized teeth, viviparous, bipedal stance

2 ©VCAA 2018 Q38 (adapted) Consider the evolution of hominins. Which one of the following statements about hominin evolution is correct?

- **A** *Homo sapiens* and *Homo neanderthalensis* are the only present-day hominin species.
- **B** Members of the *Australopithecus* genus are not classified as hominins.
- **C** *Homo habilis* was a quadrupedal primate.
- **D** All hominins are also hominoids.

3 ©VCAA 2018 Q39 (adapted) Which general trend is shown by hominin fossils?

- **A** The more recent the fossil, the more central the position of the foramen magnum in the skull.
- **B** The older the fossil, the larger the brain case that surrounds the cerebral cortex.
- **C** The more recent the fossil, the less bowl-shaped the pelvis.
- **D** The older the fossil, the smaller the jaw bones.

4 ©VCAA 2017 Q35 (adapted) Modern African *Homo sapiens* do not contain Neanderthal DNA. Modern non-African *Homo sapiens* contain a small amount of Neanderthal DNA because of interbreeding between Neanderthals and *Homo sapiens*. This interbreeding is thought to have occurred 65 000–47 000 years ago. A recent study has found *Homo sapiens* DNA in the genomes of 100 000-year-old Neanderthal remains.

From this new discovery, it would be reasonable to conclude that:

- **A** Neanderthals are the ancestors of modern Africans.
- **B** Neanderthals and *Homo sapiens* interbred in Africa more than 100 000 years ago, after which the Neanderthals spread to the rest of the world.
- **C** the ancestors of modern Africans migrated from Europe to Africa between 65 000 and 47 000 years ago.
- **D** about 100 000 years ago, Neanderthals and *Homo sapiens* interbred outside of Africa.

5 ©VCAA 2017 Q36 (adapted) In recent years, scientists have discovered that Neanderthals took care of their elderly relatives, used burial rituals for their dead and gave symbolic meaning to natural objects. Scientists have also shown that Neanderthals used complex methods to make sharp stone implements and produce glues to attach sharp stones to spears.

These discoveries suggest that Neanderthals:

- **A** traded goods with other tribes for money
- **B** cooked their food using fire
- **C** lived a social lifestyle with a highly developed culture
- **D** used a written language and rock art.

6 In 2015, the incomplete, fossilised skeletons of three individuals were found in a cave on the island of Luzon in the Philippines. The find comprised toe bones, teeth, finger bones and a femur shaft. These bones have been dated to about 67 000–50 000 years ago. The small-bodied hominin is thought to have been no taller than 1.2 metres and has been named *Homo luzonensis*.

Some of the fossils found were curved toe and finger bones, suggesting that *Homo luzonensis* may have been able to climb trees.

a What types of fossil would scientists need to find to definitively reach this conclusion? 1 mark

_____

b Some of the fossils have been dated to around 67 000 years old. Name one other human lineage that would have been around at that same time. 1 mark

_____

c *Homo luzonensis* was bipedal. State two skeletal features that would enable a bipedal gait. 2 marks

_____

_____

d Animal bones found with the bones of *Homo luzonensis* show signs of cut marks. What further information does this suggest about *Homo luzonensis*? 1 mark

_____

Answers

Chapter 1 Designing and conducting a scientific investigation, p. 2

Remember p.2

1 Research question or hypothesis
2 **a** Amount of light
   b Height of plant in cms
   c Answers will vary. Same amount of water, same type of soil, same-sized pot
   d The plants kept in the light will be taller than the plants kept in the dark.
3 Safety and ethics

1.1 Investigation design p.3

1.1.1 Observation p.3

1 Acquisition of information through your senses; sight, smell, taste, hearing and touch
2 Answers will vary. Worms are large and small, soil is different colours, white roots throughout soil
3 Framed as a question, specific, includes scientific terminology, mentions dependent and independent variables, can be answered by an investigation
4&5 Answers will vary. Are there worms of different sizes (dependent variable) because there are different nutrients in the soil (independent variable)?
6 A prediction or explanation based on an existing model or theory or observation
7 Framed as a prediction or explanation, specific, includes scientific terminology, mentions dependent and independent variables, predicts or explains results, can be supported or refuted (not proven) by an investigation
8 Answers will vary. If there are more nutrients in the soil, then the worms will grow to a larger size in cms.

1.1.2 Designing an investigation p.5

1 Answers will vary. To determine the optimal concentration of fertiliser for fruit production in tomato plants.
2 Answers will vary. What effect does fertiliser concentration have on fruit production in tomato plants?
3 Answers will vary. If the fertiliser is more concentrated, then more fruit will be produced by the tomato plant.
4 The concentration of the fertiliser

5 The number of tomato fruit produced
6 Examples: amount of light – keep plants in same area or use identical artificial lights; amount of water – give plants the same amount of water at the same time of day; nutrients in soil – plant all in the same volume and same type of potting mix; air temperature – keep all plants in the same area; age of plant – ensure all plants are the same age; size of plant – ensure all plants are the same height, have the same number of leaves
7 To allow comparison of fruit produced without fertiliser to the amount of fruit produced with fertiliser. This ensures that the DV (size of plant) is due to the IV (amount of fertiliser) and not another variable.
8 The control plant should be exposed to identical conditions as the rest of the tomato plants (same soil, water, light etc.) but will not be given any fertiliser.
9 Answers will vary depending on the hypothesis. The plants given the highest concentration of fertiliser will produce the largest number of tomato fruit.
10 Answers will vary depending on the hypothesis. The plants given the highest concentration of fertiliser do not produce the most fruit.

1.1.3 Methodologies p.6

Table 1.1: field work, controlled experiment, case study, correlational study, product process or system development, simulation, classification and identification, literature review, modelling

1.1.4 Designing your investigation to test your hypothesis p.7

1 Controlled experiment – investigating the relationship between an independent variable, fertiliser concentration, and dependent variable, number of tomato fruit
2 **a** The number of tomato fruit on each plant
   b Quantitative
   c **i**   Ensure all variables are controlled except for the independent variable (fertiliser concentration)
      ii Count the fruit on each plant more than once
      iii Follow a systematic method in counting fruit, count fruit on each plant more than once

d Answers will vary. Data could be collected once at the end of a set time period, or daily or weekly over a set time period. Experiment length one to two months

e Writing equipment or computer to record data

f Number of fruits

g Miscounting number of fruits, personal errors

3 Answers will vary. Seven identical tomato plants, seven identical pots, potting mix, 1×50 ml measuring cylinder, 1×250 ml measuring cylinder, fertiliser, trowel, seven grow lights

4 Answers will vary. Ingesting fertiliser – wear mask and gloves, wash hands; high concentration of fertiliser entering water systems – ensure excess fertiliser disposed of correctly; getting fertiliser in eyes – wear safety glasses

5 Answers will vary. The method should include details of how the independent variable will be changed, how other variables will be controlled and explain how the dependent variable will be measured. Example: plant each of the tomato plants in a pot and fill with potting mix to the same level on each pot; label the pots 1 to 7; give plant 1: 250 ml of water, give plants 2 and 3: 10 ml of fertiliser and 240 mls of water, give plants 4 and 5: 20 mls of fertiliser and 230 mls of water, give plants 6 and 7: 30 mls of fertiliser and 220 ml of water; give the plants 250 ml of water each day; give plants 2–7 the same amount of fertiliser and water as described above once a week; count the number of tomato plants on each plant after 4 weeks; repeat experiment three times.

1.1.5 Primary data p.9

1

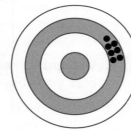

a Low accuracy, low precision   b Low accuracy, high precision

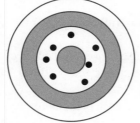
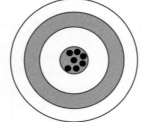

c High accuracy, low precision   d High accuracy, high precision

2 Group 1: inaccurate, precise; Group 2: inaccurate, low precision; Group 3: accurate, low precision

3 a Use a different thermometer, ensure they are using the thermometer correctly e.g. allow long enough for the thermometer to reach thermal equilibrium with the water before taking the reading, ensure the thermometer bulb is totally immersed.

b Ensure the thermometer is vertical and their eye is level with the top of the liquid in the thermometer, wait the same time period before taking a reading.

1.1.6 Minimising error p.10

1 a Due to investigator's mistakes or miscalculations. Make sure you follow the steps in the method carefully. Repeat the experiment more than once.

b Due to faults in the equipment used to take measurements. Make sure measuring equipment is calibrated and tested. Repeat measurements.

c Due to unpredictable variations in measurement process. Repeat measurements, make sure you are following the same measurement technique

d Due to the person conducting the experiment or taking the measurements having expectations of the results. Ensure all variables are controlled except for the independent variable, use double-blinded studies

2 Results that do not match current scientific understanding, presence of outliers or unexpected results, when you repeat the experiment you get varying results.

3 a systematic error
b bias
c personal error
d random error

1.1.7 Ethical guidelines p.11

Scenario 2 answers will vary.

Step 1 – Should gene therapy be used to correct genetic mutations in IVF embryos?

Step 2 – Consequence based: Gene dreams wish to screen embryos for genetic mutations and use gene therapy to correct the mutation, so the child is not born with a disability

Step 3 – Answers will vary. Beneficence: preventing children being born with disabilities; Non-maleficence: the company claims a 99% success rate and prevents destruction of embryos; Respect: the embryo/future person was not able to consent to gene therapy; Justice: is this therapy affordable for all; Integrity: is the company correctly reporting the success rate and risks of the procedure

Scenario 3 answers will vary.

Step 1 – Should individuals be required to inform their employer of genetic conditions?

Step 2 – Virtue based: is it the right thing for Grace to inform her boss about her condition?

Step 3 – Respect: does Mr Quandary have the right to communicate Grace's results without her permission? Justice: will Grace be discriminated against and lose her job? Non-maleficence: would the harm to Grace of losing her job be greater than the risk of her continuing to drive the bus?

5 **a** Scatter plot/line graph – this will show if there is a relationship between the two variables.

b

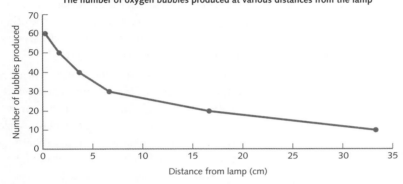

The number of oxygen bubbles produced at various distances from the lamp

c There is an inverse exponential relationship – as the distance from the lamp increases, the number of bubbles of oxygen produced decreases.

d 16–17 bubbles

6 A control would have been a plant that was not exposed to light; it is important to show that it is the light that is causing the oxygen bubbles to be produced and not something else.

7 Answers will vary. Results: it was found that the rate of oxygen production decreased as the distance of the plant from the light increased. These results can be explained because light is needed for the light-dependent stage of photosynthesis that produces oxygen (gas) as an output. Significance to scientific understanding: the rate of oxygen production/light-dependent reactions show an inverse exponential relationship, meaning that at greater distances from the light the decrease in the rate of the light reaction is less. Limitations: there is no control, results could be affected by heat from the lamp; only three trials; cannot be sure the data is reproducible; accuracy of measuring/miscounting bubbles; data from each trial is not precise. Future research: different colour light.

8 Answers will vary. The hypothesis '…' was supported. The rate of oxygen production and photosynthesis in the plant decreased with an increased distance from the light source.

Further investigations could include comparing the rate of photosynthesis with different light frequency (colour).

1.2 Scientific evidence p.15

1.2.1 Analysing your data p.15

1 Answers will vary. If the light intensity is high then the rate of photosynthesis will be higher, and more oxygen bubbles will be produced.

2 The light intensity/distance of plant from the light

3 Number of oxygen bubbles

4 33.33, 16.67, 6.67, 3.67, 1.67, 0.33

1.4 Chapter review p.19

1.4.1 Key terms p.19

Order of terms in the table to match definitions: repeatability, methodology, accuracy, method, dependent variable, independent variable, personal errors, precision, systemic errors, validity, true value, hypothesis, random errors, repeatability, qualitative data, uncertainty, outlier, quantitative data

1.4.2 Exam practice p.20

1 B
2 B
3 D
4 C
5 D
6 A

Chapter 2 The relationship between nucleic acids and proteins p.22

Remember p.22

1 Polypeptides (proteins)
2 Amino acids
3 Any two of: structural, regulatory, communication, transport, defence against pathogens, cell motility
4 a Nucleus, mitochondria, chloroplast
   b cytosol or cytoplasm
   c cytosol or cytoplasm and on rough endoplasmic reticulum
   d cytosol or cytoplasm

2.1 Nucleic acids p.23

2.1.1 Structure of nucleic acids p.23

1

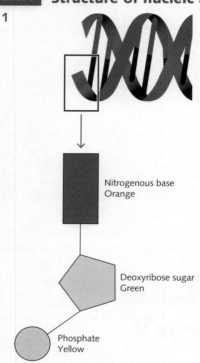

Nitrogenous base
Orange

Deoxyribose sugar
Green

Phosphate
Yellow

2 G – Guanine, C – Cytosine, T – Thymine, A – Adenine
3 The bases
4 Hydrogen bonds
5 Sugar and phosphates
6 Phosphodiester bonds

2.1.2 Structure of DNA: building a model p.24

1 Deoxyribonucleic acid
2 a 3′–TAGGCCAT–5′
   b Complementary base pairing
3 a Yes
   b Adenine and guanine have a double-ring and are therefore wider whereas thymine and cytosine have a single ring and are therefore smaller.

c The complementary base pairing rule means that cytosine (single ring) always bonds to guanine (double ring) and thymine (single ring) always bonds with adenine (double ring). In this way the distance from one strand of DNA to the other strand is consistent.

d As the bases are different sizes, the strands would be closer together where C bonded with C, and T bonded with T, and further apart where G bonded with G, and A bonded with A.

4 One strand starts with the 5′ carbon on the deoxyribose sugar pointing upwards, the other strand has the 3′ carbon facing upwards; one strand runs 5′ to 3′ and the other strand 3′ to 5′.

5 Answers will vary. Clearly shows base pairing rules, antiparallel nature of the two strands; shows the molecule is formed from many repeating units (nucleotides); shows the orientation of the nucleotides in the molecule; shows the sugar–phosphate backbone; shows the relationship between the two strands; shows the different size of the bases.

6 Answers will vary. Does not show double helix structure, does not show details of bonding between the components.

7 Answers will vary. Enables it to twist to show double helix structure, show information about bonding between the components.

8 Answers will vary. Antiparallel nature, the difference between the coding and template strand, base-pairing rules, how the base-pairing rules help determine the structure of the molecule.

2.1.3 Structure of RNA p.29

Table 2.1 Features of DNA and RNA

	DNA molecule	RNA molecule
Is it made up of nucleotides?	Yes	Yes
What is the type of sugar in its sugar-phosphate backbone?	Deoxyribose	Ribose
What are its nitrogen bases?	Cytosine, Guanine, Thymine, Adenine	Cytosine, Guanine, Uracil, Adenine
Where is it located in eukaryotic cells?	Nucleus, Mitochondria, Chloroplasts	Nucleus, Cytosol
Is it a stable molecule?	Yes	No
How many strands in each molecule?	2	1

9780170452618

1. DNA and RNA are both composed of nucleotides. The nucleotides in both differ in that DNA has the sugar deoxyribose and RNA has the sugar ribose. Their nucleotides contain one of four different nitrogenous bases: DNA – Guanine, Cytosine, Adenine and Thymine; and RNA – Guanine, Cytosine, Adenine and Uracil. They can both be found in the nucleus of the cell, but RNA is also found in the cytosol and DNA is also found in mitochondria and chloroplasts. DNA is a stable molecule whereas RNA is unstable. DNA is a double-stranded molecule held together by hydrogen bonds and twisted into a helical shape. RNA is a single-stranded molecule.

2. Reading down each column – mRNA: Messenger RNA; nucleus and cytosol; carries a copy of the DNA code from the nucleus to the ribosome to provide instructions for protein synthesis; it is a messenger – it carries a message between two entities; linear.

   tRNA: Transfer RNA; cytosol; carries specific amino acids to the ribosome needed for protein synthesis by attaching its anticodon to the complementary mRNA codon; it transfers – moves something from one place to another; clover shaped.

   rRNA: Ribosomal RNA; cytosol; combines with proteins to form the ribosome; it is ribosomal – it forms ribosomes; spherical shaped.

2.2 Gene expression p.30

2.2.1 Transcription p.30

1. Nucleus
2. Carries a copy of the DNA code out of the nucleus to the ribosome
   Plays a role in translation providing coded instructions for protein synthesis for specific genes
3. **a** Template 3′–TAGGCCAT–5′
      mRNA 5′–AUCCGGUA–5′
   b The base sequence in coding strand is identical to the base sequence in mRNA, except mRNA contains uracil in place of thymine.
4. The non-template or coding strand contains the base sequence that forms the coded instructions that determine the order of amino acids in a protein, it is not used as the template to produce mRNA (hence non-template).
5. RNA polymerase
6. Any two of: DNA is double stranded whereas pre-mRNA is single stranded; DNA has thymine as one of its bases whereas pre-mRNA contains uracil; DNA contains deoxyribose sugar whereas pre-mRNA contains ribose sugar.

7. Answers will vary. It shows that the base sequence in mRNA is the same as the base sequence in the non-template strand, it demonstrates that the template strand is used to make the mRNA strand.
8. Answers will vary. It does not show the role of RNA polymerase; it does not show exons and introns; it does not show how RNA polymerase attaches to the promoter region; it does not show the direction of transcription; it does not show the sequence of steps in the process of transcription.
9. Answers will vary. Include some steps in the model between the DNA molecule and the pre-mRNA, include arrows to show the direction of transcription, include RNA polymerase.
10. Answers will vary. The template strand is used to build the complementary pre-mRNA; the direction that transcription occurs in; the base sequence in the pre-mRNA is the same as the base sequence in the non-template strand of DNA except for U instead of A; why the template strand is used for transcription.

DNA and pre-mRNA models p.32

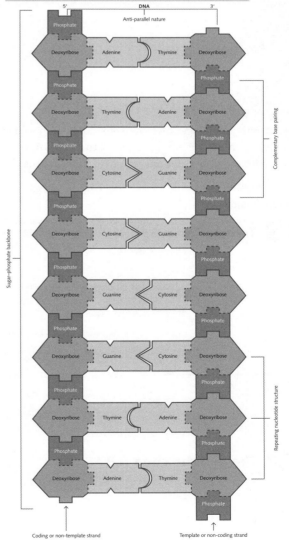

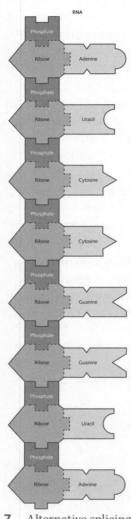

1 Introns, exons

2

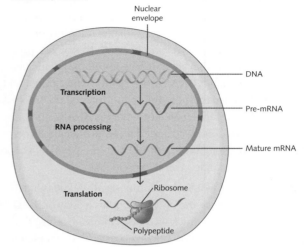

3 Transcription, RNA processing

4 1 – Promoter region signals the start of a gene, 2 – RNA polymerase positioned onto the DNA, 3 – DNA unwinds, 4 – DNA unzips, 5 – Strand of pre-mRNA built, 6 – Introns removed, 7 – Exons spliced together, 8 – Poly-A tail added/5 'methyl cap added, 9 – Mature mRNA ready to leave the nucleus

5 a GAC CUA AGG UAC
  b AAG GCU CAG UGU

6 a AUC GCU AUC GUU
  b UUA UAU CCU AUG GUG

7 Alternative splicing: different exons can be removed as well as introns, remaining exons can be spliced together in different order. This will produce mature mRNA with different base sequences. This enables an increase variety of proteins and thus reactions that can occur in cells.

8 Ribosomes

9 Ribosomal RNA/rRNA

2.2.3 **Translation p.36**

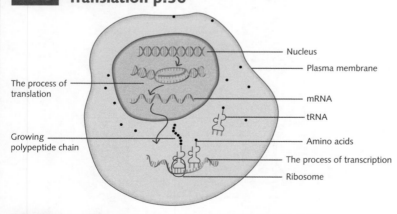

2.2.4 **The genetic code p.37**

1 a AAU
  b UGC
  c GGG

2 a AAT
  b CGC
  c GCG

3 a TTA
  b GCG
  c CGC

Activity: build a model of translation

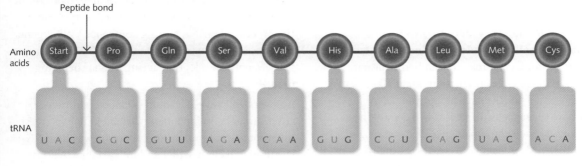

Peptide bond

Amino acids: Start — Pro — Gln — Ser — Val — His — Ala — Leu — Met — Cys

tRNA: U A C | G G C | G U U | A G A | C A A | G U G | C G U | G A G | U A C | A C A

mRNA: A U G | C C G | C A A | U C U | G U U | C A C | G C A | C U C | A U G | U G U

1 Cytosol at the ribosomes; either free floating in the cytoplasm or attached to the endoplasmic reticulum

2 Codon

3 AUG is the codon that initiates or starts protein synthesis. UAA, UAG and UGA are all stop codons – they terminate or stop polypeptide synthesis.

4 a The code is degenerate, most amino acids have two or more codes; if a mutation occurs in the DNA, there is less chance that it will change the amino acid coded for and the protein will still contain the same amino acid sequence and still be functional.

   b Met-Pro-Gln-Ser-Val-His-Ala-Leu-Met-Cys

   c i The amino acid would be His instead of Gln, this would alter the structure of the protein and it would be non-functional.

      ii Both codes specify Ala; that is, there is no change to the polypeptide chain.

      iii UGA is a stop codon, this would mean Arg would not be added and the polypeptide chain would be shorter, the protein would not function.

5 Answers will vary. Strengths: easy to see anticodon and codon complementary pairing; shows the anticodon determines the amino acid carried by the tRNA; shows the growing polypeptide chain; shows tRNA can be loaded and unloaded with specific amino acids; shows that the codons in the mRNA determine the amino acid sequence in the polypeptide by matching with corresponding tRNA anticodons. Limitations: does not show true shape of tRNA; the ribosome is not included; it is static and does not show how the ribosome moves along the mRNA does not show the 3D molecular shapes that the molecules have.

6 Answers will vary. Start in the middle of the mRNA strand and match three tRNA anticodons to this section of the mRNA and glue them down, with their specific amino acids attached. To the left of this, show the amino acid chain that has already

been translated and the tRNA that have already added their amino acids. Show the tRNA on the right waiting to enter the ribosome and add a diagram to represent the ribosome. Ensure to add arrows to show the direction of movement.

7 Answers will vary. The relationship between mRNA codons and tRNA anticodons and how the complementary pairing between the mRNA codons and the tRNA codons determines the specific amino acid sequence in the polypeptide chain; that the tRNA carry the correct amino acid to add to the growing polypeptide chain.

2.2.5 Scientific literacy p.43

1 Yes. The article is current, the name of the scientific institution is given along with the name of a scientist involved and there is reference to a previous study that delivered the same results. But we do not know where this article was published, and the previous study was carried out at the same scientific institution.

2 Answers will vary. 'Helpful molecules read our DNA, create short little RNA messages, and send them outside the nucleus to tell the rest of the cell which proteins need to be built.' The base sequence in DNA form coded instructions that control the production of proteins. In a process called transcription, an enzyme called RNA polymerase can create a copy of a gene's DNA sequence in the form of messenger RNA (mRNA). The mRNA is able to leave the nucleus and carry its copy of the DNA base sequence to the ribosomes in the cytosol, the ribosomes can read the coded instructions in the mRNA and use it to synthesise a protein by the process of translation.

3 RNA processing

4 RNA is a copy of the DNA base sequence found in one part of the DNA molecule, if the RNA copy is damaged further copies can be made using the DNA master copy that is safely stored in the nucleus.

5 Answers will vary. The genes you inherit from your parents do not change during your lifetime. Genes are sections of a large molecule called DNA; this is stored in the nucleus inside all of your cells. DNA carries instructions to make molecules called proteins that perform many important roles in your body. The part of your cell that reads the instructions and makes the proteins is outside the nucleus, but your DNA can not leave the nucleus. To solve this, a copy of the instructions is made, the copies are called messenger RNA or mRNA. Messenger RNA carries the instructions to the protein factories (ribosomes) so that they can be used to build proteins. We thought that mRNA could not be changed once it left the nucleus, but scientists have found that this is not correct. They have found that in the nerve cells of squid, messenger RNA was changed once it left the nucleus.

6 Answers will vary. What internal/external factors lead to mRNA editing? Is all mRNA able to be edited or only mRNA transcribed from some genes? How is the process of mRNA editing controlled? What gene or genes are important for enabling mRNA editing? Is there natural selection acting upon processing of mRNA?

7 Answers will vary. If a squid is placed in different environments (independent variable), such as different water temperature, then the way its mRNA is edited will change (dependant variable). A study could keep genetically identical squid of the same species, health, age in different water temperatures; compare the mRNA or proteins found in their nerve cells.

2.3 Gene regulation p.45

2.3.1 The *trp* operon: an example of gene regulation – part A p.45

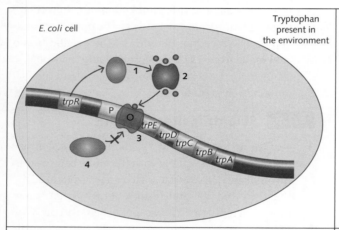

If tryptophan is present it binds to the *trp* repressor (1) that is coded for by the *trp*R regulatory gene. The repressor is activated (2), changing its shape allowing it to bind to the operator (3). This stops RNA polymerase (4) from being able to bind to the promoter and preventing the transcription and translation of the genes in the *trp* operon. No tryptophan is produced as it is already present in the cell.

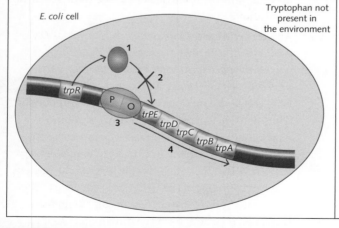

If no tryptophan is present the *trp* repressor (1) that is coded for by the *trp*R regulatory gene cannot be activated. It cannot bind to the operator (2); RNA polymerase (3) binds to the promoter, allowing transcription (4) and translation of the genes in the *trp* operon. Tryptophan is produced as it is needed by the cell.

2.3.2 The *trp* operon: an example of gene regulation - part B p.46

2 a D
b C
c C
d B
e A
f D
g C
h C
i C
j D

2.4 Proteins p.46

Activity: build a model to show the four levels of protein structure p.46

1   Endoplasmic reticulum and ribosome

2   **a**   Peptide bonds

    b   Hydrogen bonds

3   Tertiary: enzymes have an active site. This is a complementary shape for a specific substrate, if the 3D shape of an enzyme is altered, the active site shape will change, it will no longer be complementary to its substrate, the substrate will not be able to enter the active site and the enzyme will not be able to catalyse the reaction involving that substrate.

4   Mistakes in translation will alter the primary structure (order of amino acids) in the polypeptide chains that form the enzyme. This is turn will alter the secondary structure, the tertiary structure and the quaternary structure of the enzyme, the enzyme will be a different 3D shape and the active site will be altered. The enzyme will not be able to bind to its substrate and catalyse a reaction that is important to the correct functioning of the cell.

5   Strengths: shows physical structure of each level; shows that one level of structure determines the next level; shows that more than one polypeptide chain is needed to form the quaternary structure; shows the specific order of amino acid and how the R-group interactions determines the 3D shape. Limitations: does not show the different bonding; does not show why the protein folds in a particular way; does not show that one polypeptide chain could have both helices and pleated sheets; does not show the 3D structure.

6   Answers will vary. Add rules for folding the tertiary structure, use different materials to represent different bonding at different levels.

7   Answers will vary (must relate to the difference between each level). Primary structure is linear, that each level depends on the previous level and the R-group interactions determines the 3D shape of the protein.

2.5 The protein secretory pathway p.51

2.5.1   Telling the story of the protein secretory pathway p.51

Step 1

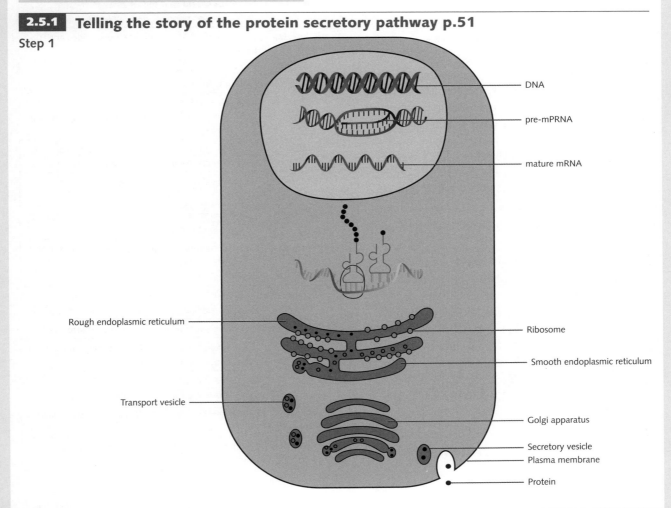

Step 2 *Transcription* is the process of making an RNA copy of a gene sequence. This copy is called a pre-mRNA. The pre-mRNA is processed to become mature mRNA – introns are removed, exons are spliced together, Poly-A tail and Guanine cap is added, alternative splicing may occur. The mature mRNA leaves the nucleus and travels to the cytosol and attaches to a ribosome.

Translation: the code on the mRNA is read at the ribosomes; ribosomes hold the mRNA in place while tRNA molecules carry in specific amino acid coded for by the mRNA. The amino acid is joined to a growing chain to form a polypeptide. The complementary tRNA anticodon binds to the mRNA codon.

Protein secretory pathway: proteins that are secreted from the cell are made by ribosomes embedded in the rough endoplasmic reticulum; the synthesised polypeptide chains move into the inside of the rough endoplasmic reticulum. They are folded correctly to form specific 3D shapes; they are packaged into small membrane-bound packages called transport vesicles and then they travel to the Golgi apparatus. The proteins are further modified in the Golgi apparatus to become functional proteins before they are packaged in secretory vesicles. They move to the plasma membrane of the cell and they fuse with the plasma membrane that opens to release them out of the cell.

1   Proteins that remain in the cell are made at free ribosomes floating in the cytoplasm and could be involved in cell motility, e.g. form microtubules to move flagella, or catalyse reactions within the cell. For example, enzymes or transport materials, haemoglobin carries oxygen in red blood cells, proteins that form channels in plasma membranes. Proteins that leave the cell have roles including carrying messages between cells (e.g. hormones) and defending against foreign organisms (e.g. antibodies).

2   The Golgi apparatus consists of stacks of flattened membrane-enclosed and fluid-filled saccules (cisternae). The endoplasmic reticulum also comprises an extensive network of membrane-enclosed sacs and tubules. A difference is that the endoplasmic reticulum is continuous with the nucleus; and the Golgi apparatus is separate to the nucleus. Some endoplasmic reticulum has ribosomes embedded in it, rough·ER, whereas the Golgi apparatus does not.

3   They would both be composed of a phospholipid bilayer.

2.6 Chapter review p.53

2.6.1 Find the errors p.53

1   a   Incorrect labels – swap base and phosphate
    b   All base pairs are incorrect should be A–T and C–G
    c   Switch the coding and template strand labels and swap 5′ and 3′ labels on DNA strands
    d   The tRNA are carrying the incorrect amino acids: ACC should be Thr, UUG should be Asn, CUA should carry Asp, UUA should carry Leu.

2.6.2 Key terms p.53

Key terms	Definition
DNA	an information molecule that is the basis of an organism's genetic material
Anticodon	the three nucleotides in tRNA that bind to the complementary codon in mRNA
Antiparallel	parallel but orientated in opposite directions
mRNA	RNA template copied from DNA that takes instructions for polypeptide synthesis from the nucleus to the cytoplasm
Cisternae	a flattened membrane disc that makes up the Golgi apparatus and endoplasmic reticulum
rRNA	an RNA strand that serves as a structural component of a ribosome
Poly-A tail	a chain of 100–200 adenine nucleotides added at the 3' end of an mRNA strand
Template strand	a strand of DNA that is copied during DNA or RNA synthesis
Coding strand	the DNA strand that has the same sequence of nucleotides as the mRNA (except it has T instead of U)
tRNA	an RNA molecule that transports an amino acid to the ribosome for assembly into a polypeptide
Peptide bond	a chemical bond that links two amino acids in a chain
Regulatory gene	a gene whose product switches on or switches off expression of one or more other genes
Transport vesicle	a small membrane-bound sac containing protein that is transported from the Golgi apparatus to the plasma membrane for release
Degenerate	a property of the genetic code in which most amino acids are encoded by two or more codons
Transcription	the process by which DNA is copied into RNA
Exon	a segment of DNA or RNA containing information that codes for a polypeptide
Intron	a segment of DNA or pre-mRNA that does not code for a polypeptide
RNA polymerase	the enzyme that catalyses the synthesis of RNA
Amino acid	a nitrogen-containing compound that is the monomer from which proteins are built

(continued)

Key terms	Definition
Alternative splicing	exons are removed with the introns to produce mRNA molecules of different length and sequence
Translation	the process of turning the nucleotide sequence of mRNA into the amino acid sequence of a polypeptide

Key terms	Definition
Ribosomes	a small structure consisting of RNA and proteins where amino acids are joined to form polypeptides
Polypeptide	a linear polymer built from amino acid monomers

Chapter 3 DNA manipulation techniques and applications p.58

Remember p.58

1 DNA is a polymer containing many repeating units called nucleotides. A DNA nucleotide consists of deoxyribose sugar, a phosphate group and one of four nitrogenous bases: adenine, cytosine, guanine and thymine. A DNA molecule consists of two strands of nucleotides held together by hydrogen bonding between complementary bases on each strand: adenine with thymine and guanine with cytosine. The base pairs are like rungs of a ladder, the sugar and phosphate groups form the backbone of the molecule, the two strands run in opposite directions to each other and twist to form a double helix.

2 Transcription, RNA processing, translation

3 Enzymes catalyse or speed up by lowering the activation energy of the rate of every chemical reaction that occurs in living things. Without enzymes, reactions would occur too slowly to sustain life.

4 Mitosis and meiosis

3.1 Genetically modified organisms p.59

Answers will vary. For example:

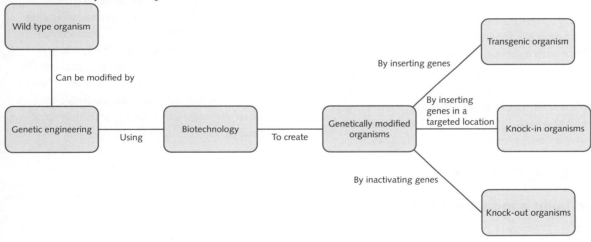

3.2 Enzymes for modifying DNA p.59

1&2

Table 3.1

Enzyme	After cutting	Blunt or sticky end?
Acc161	5'TGC GCA3' 3'ACG CGT5'	Blunt ends
Hac1	5' GATC 3' 3' CTAG 5'	Sticky ends
Gst1	5' G GATCC3' 3' CCTAG G5'	Sticky ends
Pac1	5'TTAAT TAA3' 3'AAT TAATT	Sticky ends

3 Sticky ends: two fragments with complementary overhanging ends will join together by complementary base pairing and create a more stable molecule.

3.3 CRISPR-Cas9 p.60

1 Answers will vary. To investigate the accuracy of CRISPR-Cas9 when used to edit a specific gene locus; or to determine the error rate of CRISPR-Cas9 when used in gene editing; or to determine that the CRISPR-Cas9 system can be used to edit a gene.

2 a No. Mice could be used instead of humans.

   b A positive control is a control group in an experiment that uses a treatment that is known to produce results. In this case, a modified gene would express the desired protein.

c Somatic therapy, as it will produce results faster. Germline therapy: editing the genome of sperm and egg cells; the new gene would be present in all cells of the organism following meiosis and fertilisation. Somatic therapy: editing body cells of an organism. In somatic gene therapy, only the targeted cells will be edited, and the effect of the new gene will be seen when the cell divides (mitosis).

d Check for the expression of modified genes. If the desired protein is produced or the phenotype is visible.

3 Answers will vary. Potential for mutation – use form of CRISPR-Cas9 known to be accurate.

Escape of modified organisms to environment – ensure strict protocols to prevent this.

4 a Answers will vary. Beneficence: what are the consequences for the child if gene therapy does not occur? What are the potential benefits in terms of quality of life, lifespan etc? Is it possible to treat this disorder when the child is old enough to have a say? Non-maleficence: what are the chances the gene therapy could cause harm? If it is possible to correct the genetic mutation is it ethical to allow the child to be born with a genetic disorder?

b Answers will vary. Beneficence: if the treatment dramatically improves the lifespan/quality of life of individuals, should it not be offered at all if only the wealthy can afford it? How many people would benefit from this therapy? Is it better to focus on development of more affordable therapies? Non-maleficence: the cost of the therapy further harms a group of people who cannot access the treatment. Would this mean more affordable treatments for this condition would no longer be developed? If the treatment is available is it ethical to prevent a part of the population accessing it?

c Answers will vary. Beneficence: how safe are these therapies and are they worth the risk? Would they provide substantial benefits to the individual / improve quality of life? If the therapies are proven safe why not give your child characteristics that will lead to benefits for them? Non-maleficence: risk introducing potential harmful mutations for no real health benefit, could lead to unreasonable expectations on the child to achieve because they have the gene.

d Answers will vary. No, as edits will be passed through generations to individuals who did not give consent. Yes, edits will overcome disadvantageous mutations.

e Answers will vary. Social: increased life span and quality of life, equity of access, changing expectations of what medical intervention can achieve. Economic: reduced health care costs, increased work capacity for individuals. Legal: privacy of information, how will the use of gene therapy be regulated. Political: can interest groups influence legislation around gene therapy?

3.4 Amplifying DNA p.63

Activity: build a model of the polymerase chain reaction

1 Denaturation

Strands heated to 95°C to separate strands

2 Annealing

Cooled to 50–60°C to allow primers to anneal

Reverse primer

Forward promer

3 Extension

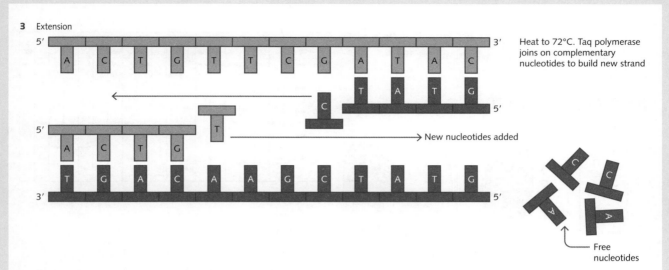

Heat to 72°C. Taq polymerase joins on complementary nucleotides to build new strand

New nucleotides added

Free nucleotides

1 Answers will vary. Taq polymerase is not shown, the thermocycler is not included, the model only shows one cycle, does not show the buffer solution,

2 Answers will vary. Show the results after a number of cycles or Taq polymerase at the third stage of annealing.

3 Answers will vary. That the primers attach to the 3′ ends of the target DNA, the order of the steps involved, what happens at each temperature, that all materials are in the one tube at the start of the experiment and by changing the temperature of the reaction allows each step to proceed.

3.5 Gel electrophoresis p.67

Table 3.2 Answers may vary slightly due to ruler measurements.

HindIII		EcoRI		BamHI	
Distance (mm)	Length (bp)	Distance (mm)	Length (bp)	Distance (mm)	Length (bp)
60	23 130	70	21 226	80	16 841
120	9416	140	8000	110	12 740
160	6557	160	6557	150	7280
200	4361	170	5900	170	5900
280	2322	250	5200	180	5450
310	2027				

1 47 813 bp

2 Do not match. Inaccuracy in measurement, inaccuracy in calculating length in base pairs, more than one fragment of the same size produced

3 As a control to show the size of the uncut λ-DNA

4 It was the longest piece of DNA having the largest number of base pairs; therefore, it did not move that far through the agarose gel.

5 Gel electrophoresis separates DNA molecules by size. The negatively charged DNA fragments are pulled through a gel matrix towards a positive electrode. Smaller-sized DNA fragments will travel more quickly and further than larger DNA fragments. The position of different-sized fragments in the gel can be shown with a binding dye and a standard of

known DNA fragment sizes can be used to estimate the size of the unknown DNA fragments. The viscosity of the gel will also affect the rate at which the DNA fragments move through the gel. The great the density of agarose in the gel the slower the DNA fragments will move.

6 Restriction enzymes are used to cut DNA molecules into smaller pieces; the enzymes cut the DNA molecule at specific recognition sequences in the DNA. The number and size of fragments produced will depend on the number and position of restriction sites in the DNA sequence.

7 Answers will vary. Each of the three restriction enzymes produced different numbers and sizes of fragments when used to cut the λ-DNA. There are five restriction sites for HindIII producing DNA six fragments: four in EcoRI producing five DNA fragments and five in BamHI producing five DNA fragments.

3.6 DNA profiling p.70

1

	1		2		3		4		5		6
A	T	C	G	A	T	A	T	C	G	A	T
C	G	T	A	A	T	A	T	A	T	A	T
C	G	C	G	G	C	G	C	T	A	G	C
C	G	T	A	C	G	C	G	A	T	C	G
G	C	A	T	T	A	A	T	A	T	T	A
G	C	A	T	T	A	G	C	G	C	T	A
A	T	G	C	C	G	G	C	C	G	G	C
T	A	A	T	C	G	A	T	T	A	C	G
C	G	A	T	A	T	A	T	T	A	C	G
C	G	T	A	T	A	T	A	G	C	C	G
G	C	T	A	G	C	T	A	A	T	C	G
T	A	C	G	T	A	C	G	G	C	G	C
G	C	A	T	G	C	G	C	G	C	G	C
T	A	G	C	G	C	A	T	A	T	G	C
A	T	T	A	A	T	A	T	A	T	A	T
A	T	T	A	T	A	A	T	T	A	T	A
G	C	C	G	C	G	A	T	T	A	C	G
C	G	G	C	C	G	T	A	C	G	C	G
T	A	T	A	G	C	T	A	T	A	T	A
T	A	C	G	A	T	T	A	T	A	G	C
C	G	C	G	G	C	A	T	A	T	G	C

Restriction enzyme	Number of cuts	Number of fragments	Length of DNA fragments
BamHI	1	2	5,12
G C G C A T T A C G C G	0	1	21
	1	2	4,13
	0	1	21
	0	1	21
	1	2	21
HindIII	1	2	2, 15
A T A T G C C G T A T A	0	1	21
	1	2	1, 16
	0	1	21
	1	2	4, 13
	1	2	nor this one
EcoRI	0	1	21
G C A T A T T A C G	1	2	7, 10
	0	1	21
	1	2	7,10
	1	2	4,13
	0	1	21

2

3 Individual 4. This individual has identical size fragments as the mother for all three restriction enzymes.

4 Ensures that different restriction fragments are produced for individuals. Comparing the presence and location of more DNA sequences in individuals improves accuracy of results.

Scenario 2: Paternity testing p.72

The man could not be the father of the child. One of the child's restriction fragments must be inherited from each parent. Neither of the child's BamHI fragments could have been inherited from the man and neither of the child's HindIII fragments could have been inherited from the father.

Identifying errors p.72

a Contamination of DNA samples; failing to add all required reactants (such as not adding either DNA or enzyme) to all samples; not adding the correct amount of a reactant to one sample; loading differing amounts of DNA into each well; when attempting to load the sample in the well missing the well altogether

b Incorrectly using equipment, reversing the charge on the electrophoresis gel; using the wrong method to collect and store all DNA samples; thermocycler is not calibrated to correct temperature for each cycle of PCR

c Misinterpretation of results to match expectations, not questioning or repeating tests if the results match expectations

d Incorrectly using standards or using inappropriate standards to find the size of DNA fragments; not using enough STR loci to analyse the DNA samples, not using a range of restriction enzymes to digest the DNA samples; not using appropriate primers in PCR; the primers may be able to attach to more than one part of the target DNA; storing enzymes at incorrect temperatures

e Not using equipment correctly for all measurements; differing gel concentrations; changing the voltage across the gel; carefully measuring quantities; control all variables

3.7 Recombinant plasmids and human insulin p.73

3.7.1 Recombinant plasmids p.73

1. A restriction enzyme is used that will cut the plasmid once only, producing sticky ends.

   The same restriction enzyme is used to cut on either side of the gene to be inserted, as close as possible to the start and end of the gene, producing sticky ends that are complementary to the sticky ends of the cut plasmid. The gene and the plasmid are mixed together with DNA ligase, the gene will be joined into the plasmid. This creates a recombinant plasmid, a plasmid that now contains a gene from another organism.

   The recombinant plasmid is sometimes taken up by bacterial cell, the bacteria replicate, some of the new bacteria formed contain the recombinant plasmid, the plasmid replicates inside the bacterial cells leading to many copies of the inserted gene.

2. They are able to pass from one bacterium to another, they are good vectors to carry DNA or genes into bacteria. They are copied many times inside bacterial cells and are copied when bacteria replicate (binary fission). Their small size allows them to be distinguished from the main bacterial chromosome, and they are still large enough to be manipulated in the lab. They are stable and easy to engineer. They can easily be selected for using antibiotic resistance genes and agar containing the antibiotic.

3. DNA is universal, the same sequences code for the same amino acids in all organisms. If the gene is transcribed and translated it will produce the same amino acid sequence.

3.7.2 Synthesising the human insulin gene p.74

1. 2 – Obtain a plasmid. 3 – So that all fragments have complementary sticky ends. 4 – The gene will be incorporated into the plasmid genome due to their complementary sticky ends. 5 – Add DNA ligase. 6 – Add recombinant plasmids to a bacterial culture. 7 – To ensure only bacteria containing the recombinant plasmids are cultured. 8 – A colony is grown to make identical bacteria all containing the recombinant plasmid.

2. Answers will vary. Cheap to produce; insulin from animals can cause allergic reactions recombinant insulin does not cause allergic reactions; does not involve death of animals; large quantities of the product can be made easily.

   3 a Modifying structural genes would lead to the production of new or functional

proteins; improve health; allow production of therapeutics; increase yield of crops

b Able to increase the expression of a gene; increase production of a particular protein or turn off genes; stop the production of a protein; investigate the function of a particular gene.

3.7.3 Plasmid X p.75

1. To identify bacteria cells that have taken up a plasmid; both the original plasmid and the recombinant plasmid contain an ampicillin resistance gene. The bacterial cells that have not taken up the plasmid will not grow in the presence of ampicillin.

2. To disrupt the tetracycline resistance gene; the recombinant bacteria will not provide resistance to tetracycline but will grow in presence of ampicillin and therefore will not survive in the presence of tetracycline.

3.

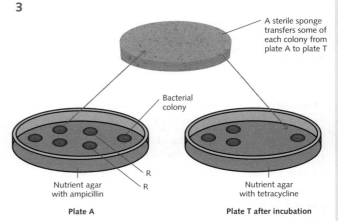

A sterile sponge transfers some of each colony from plate A to plate T

Bacterial colony

R

R

Nutrient agar with ampicillin

Nutrient agar with tetracycline

Plate A

Plate T after incubation

4. To prevent development of antibiotic resistant bacteria

5. Jelly fish gene that produces a green fluorescent protein (GFP)

3.8 Genetic engineering in agriculture p.77

3.8.1 Literature review p.77

1. Opinion piece. It is one person's view, not supported by scientific data. No scientific references for claims

2. a Correlational study
   b Case study
   c Case study/controlled experiment
   d Literature review

3. Research question – missing; Introduction – present; Methodology – missing; Results/data – missing; Discussion – present; Conclusion – present; References and Acknowledgements – missing, independent variables, dependent variables, controlled variables and control – missing

3.9 Chapter review page 79

3.9.1 Key terms page 79

Annealing	attaching primers at each end of target DNA
Blunt end	end of a DNA fragment created following cleavage by a restriction enzyme that cuts DNA at the same position on both strands
Cas9 protein	an endonuclease that cuts double-stranded DNA at a target location in the genome
Denaturation	a permanent change in the molecular structure of a protein
DNA ligase	an enzyme that catalyses the formation of a phosphodiester bond between two pieces of DNA
DNA polymerase	the enzyme that catalyses the bonding of nucleotides to form new strands of DNA
DNA profiling	comparison of individuals based on patterns of non-coding base sequences in the genome
DNA sequencing	the process of establishing the nucleotide sequence of a piece of DNA
Gel electrophoresis	a technique that separates DNA fragments according to their size and charge
Gene cloning	the process of often using plasmids and bacteria to make numerous identical copies of a gene
Knock-in organism	an organism in which DNA has been inserted into a specific locus
Polymerase chain reaction	used to copy a DNA template, making millions of the same piece of DNA
Polymorphism	a variation in DNA sequences among individuals
Recombinant plasmid	a plasmid with foreign DNA inserted into it
Restriction endonuclease	an enzyme that cuts DNA at a specific restriction site
Restriction fragment	a short fragment of DNA generated after the cutting of a longer DNA fragment by a restriction enzyme
Restriction site	a specific nucleotide sequence that is recognised as a cleaving site for a restriction enzyme
Short tandem repeat	a short non-coding region of DNA of up to five bases that is repeated many times in the genome of an organism
Sticky end	an end of a DNA fragment created following cleavage by a restriction enzyme that cuts DNA at different positions on each strand
Vector	a vehicle used to transfer DNA sequences from one organism to another

3.9.2 Exam practice p.82

1 B

2 C

3 B

4 a The gene mutations being screened for are recessive; to be able to pass the genetic disorder onto offspring both parents must carry the mutation; the child needs to inherit one copy of the mutated allele from each parent.

b PCR, polymerase chain reaction

Denaturation: heat to 95°C to break hydrogen bonds between the bases and separate the strands

Annealing: cool to 50–60°C to allow primers to join to complementary sequences on opposite ends of each strand (attach to 3′ end of each strand), the lower temperature is needed to allow hydrogen bonds to form between complementary base pairs.

Extension: heat to 72°C, optimum temperature for the enzyme Taq polymerase that starts at each primer and synthesises a new DNA strand by joining complementary nucleotides in the 5′ to 3′ direction. There are now two identical copies of double stranded DNA.

c Gel electrophoresis – separates strands of DNA according to size, negatively charged DNA fragments are pulled through a gel matrix towards a positive charge, a binding dye is attached to the DNA allowing the final position of the DNA fragments in the gel to be visualised, small fragments will move further than larger fragments, creating a banding pattern that can be used to estimate size of the DNA fragments. A known DNA standard ladder can be run in the gel to be used as a comparison and a size reference to the unknown fragments of DNA to estimate the size of the fragments.

d Answers will vary. Ethical: should they inform their child that she is a carrier; should they encourage their daughter to have children; did they make the right choice in going ahead with pregnancy; did the benefits outweigh the harms to their daughter. Social: will their daughter be discriminated against; will their daughter have increased health costs if she decides to reproduce.

Chapter 4 Enzymes and the regulation of biochemical pathways p.85

Remember p.85

1 To catalyse, or speed up, the rate of chemical reactions by lowering the activation energy required for the reaction to take place

2 To capture light energy and convert it into chemical energy in the form of glucose

3 Chloroplasts

4 To break down glucose to release energy that is used to produce molecules of ATP

5 Mitochondria

4.1 Biochemical pathways for cell metabolism p.85

4.1.1 Organising key terms p.86

1

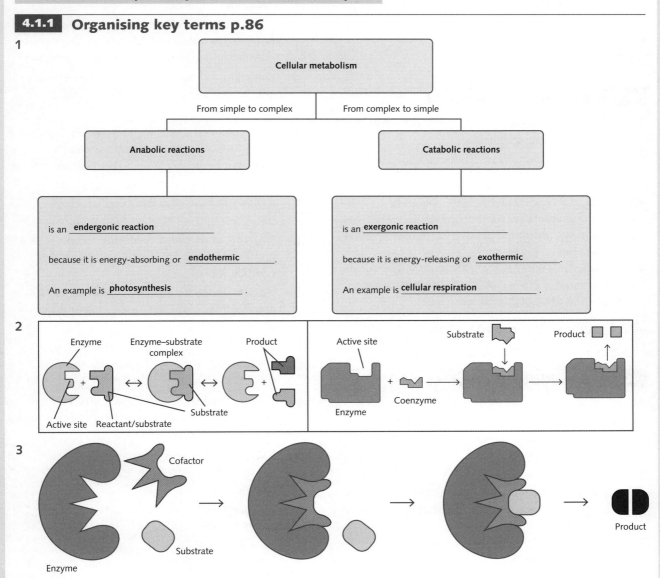

Cellular metabolism

From simple to complex — From complex to simple

Anabolic reactions

is an **endergonic reaction**

because it is energy-absorbing or **endothermic**.

An example is **photosynthesis**.

Catabolic reactions

is an **exergonic reaction**

because it is energy-releasing or **exothermic**.

An example is **cellular respiration**.

2

Enzyme — Enzyme–substrate complex — Product

Active site — Reactant/substrate — Substrate

Active site — Substrate — Product — Enzyme — Coenzyme

3

Cofactor — Substrate — Enzyme — Product

4 Cofactors (inorganic) and coenzymes (organic) are needed to activate some enzymes, cofactors alter the active site's shape and /or charge to allow the substrate to bind to the active site. Coenzymes carry chemical groups between enzymes.

4.1.2 Biochemical pathways and coenzymes p.87

1 ATP provides energy and phosphate for this reaction to occur; glycolysis could not occur without ATP.

2 Endothermic, it is using energy: ATP

3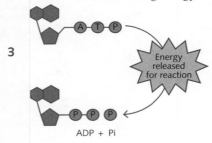

4 a Product A
  b Product B
  c Product B

4.1.3 Scientific literacy p.88

1 To construct a synthetic cell that can sustain itself and that can grow and divide

2 Arginine $\xrightarrow[\text{ADP + Pi} \quad \text{ATP}]{\text{Enzyme}}$ Ornithine (+ ammonium + carbon dioxide)

3 Presence (or concentration) of the substrate arginine, presence (or concentration) of the enzymes, production of ornithine (needed for arginine to enter the vesicles)

4 Published in a science journal *ScienceDaily*, academic institution that conducted the research is named as well as the name of the scientist.

4.2 Enzymes: the key to controlling biochemical pathways p.90

4.2.1 Specificity of enzymes p.90

1 The shape and charge of an enzyme's active site (the lock) is complementary to its specific substrate (the key). Therefore, when the substrate and enzyme bind, the reaction takes place.

2 The shape of an enzymes active site changes when it meets its specific substrate to allow the substrate to fit correctly in the active site allowing the reaction to take place.

3 In the lock-and-key model, the active site is static; its shape does not change. In the induced-fit model, the active site changes to become complementary to fit its substrate only when it encounters its substrate; the active site reverts to original shape after the products are released.

4 The induced fit better explains how the enzyme can catalyse a reaction, the changes in the active site shape stretches and bends bonds in the substrate, increasing reactivity. The lock-and-key model does not explain how this could occur.

5 Answers will vary. The cushions of a chair change shape to match your body when you sit in it; when you take someone's hand, you change the shape of your hand to match their hand etc.

4.2.2 Enzymes need help: coenzymes and cofactors p.91

1 a Loaded – carry protons, electrons or chemical groups that they are needed for anabolic reactions to occur

  b Unloaded – can accept protons, electron or chemical groups that are released from catabolic reactions

  c Coenzyme – non-protein organic substance, can be loaded or unloaded with protons, electrons or chemical groups, carry these between reactions

  d Substrate – a substance on which an enzyme acts/ a reactant for an enzyme-controlled reaction.

  e ATP – a loaded coenzyme that carries energy and a phosphate group, needed for photosynthesis, and produced by cellular respiration

  f ADP – an unloaded coenzyme, can accept phosphate and energy in cellular respiration

  g $NADP^+$ – an unloaded coenzyme accepts H^+ and electrons released in the light-dependent reaction of photosynthesis

  h NADPH – a loaded coenzyme that carries H^+ and electrons to the light-independent reactions of photosynthesis

  i NAD^+ – unloaded coenzyme accepts H^+ and electrons from glycolysis and Krebs cycle in cellular respiration

  j NADH – loaded coenzyme that carries H^+ and electrons to the electron transport chain of cellular respiration

  k FAD – unloaded coenzyme accepts H^+ and electrons during Krebs cycle of cellular respiration

  l Unloaded coenzymes are positively charged except for ADP and FAD.

2 Coenzymes are non-protein organic molecules which carry chemical groups between enzymes. Cofactors are non-protein chemical compounds that bind with an enzyme.

3 Both are necessary for the correct functioning of some enzymes. Cofactors are needed to help the enzyme form the correct shape to be able to interact and complementarily bind with its substrate; coenzymes carry protons, electrons or groups of atoms to an enzyme needed for the reaction it is catalysing.

4.3 Photosynthesis and cellular respiration p.93

4.3.1 Effect of temperature p.93

1. Answers will vary. Include how the IV will affect the DV. How does temperature effect the rate that diastase can break starch down to maltose?

2. Answers will vary, must be an explanation or prediction and include the IV and DV. *If the rate that diastase breaks down starch to maltose is highest at 37°C; then the rate of break down will reduce at temperatures above and below this.*

3. a Temperature
   b How much starch has been broken down to maltose

4. Controlled experiment: this is investigating the relationship between two variables (IV and DV)

5. No, this method has not controlled all variables except for the independent variable. A statement of how the variables will be controlled is required: *the amount of diastase added is the same, only the temperature will change.* (Methods will vary.)
   1: Set up five test tubes, to each test tube add 2 ml of diastase.
   2: Label test tubes with 0, 10, 30, 45 and 60°C.
   3: Place each test tube in a water bath, set to the temperature on the test tube, leave for 5 minutes.
   4: Add 10 ml of the same concentration starch solution to each test tube.
   5: After 5 minutes, remove from the water bath.
   6: Add five drops of iodine to each test tube and record the colour (dark blue = high starch, yellow = no starch)

6.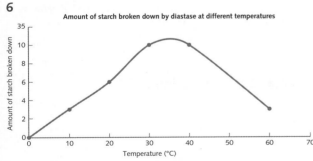
   Amount of starch broken down by diastase at different temperatures

7. High temperatures will break bonds in the protein molecule, this will alter the 3D shape of the enzyme, the active site will alter and will no longer be complementary to the substrate as the enzyme is now permanently denatured the reaction will not proceed.

4.3.2 Effect of changing pH p.96

1. Answers will vary. Do different levels of pH have the same effect on the activity of amylase, pepsin and arginase?

2. Answers will vary. If pepsin, amylase and arginase show the same optimum activity with the same pH, then altering the pH will affect the activity of each enzyme the same.

3. a pH of reaction
   b % enzyme activity

4.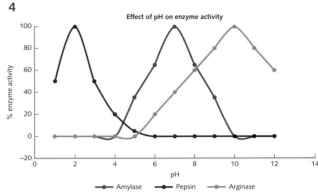
   Effect of pH on enzyme activity

5. Answers will vary based on hypothesis. The results refute the hypothesis, they show that each enzyme has a different optimum pH.

6. a Amylase – pH 7
   b Pepsin – pH 2
   c Arginase – pH 10

7. a Amylase – mouth
   b Pepsin – stomach
   c Arginase – liver

8. The hypothesis that pepsin, amylase and arginase will all show optimum activity at the same pH was not supported by the results. It was found that each enzyme was most active at a different pH: amylase at pH 7, pepsin at pH 2 and arginase at pH 10. Therefore, these three enzymes have a different optimum pH. Further investigations into the optimal pH of other enzymes would be useful.

4.3.3 Effect of substrate and enzyme concentration p.99

1. Presence of amino acids

2. Answers will vary. Will the rate of protein broken down by the enzyme, protease, be faster with a higher enzyme concentration or a higher substrate (protein) concentration?

3. Answers will vary. If a higher concentration of protease, rather than a higher concentration of substrate, produces a faster rate of protein breakdown, then more amino acids will be found when the concentration of protease is higher, not when the concentration of substrate is higher.

4. a High enzyme concentration and high protein concentration
   b The rate at which protein is broken down to amino acids

5. Same temperature, pH, time for the reaction, same-sized container

6 Protein with no enzyme added to see if protein can break down in the absence of the enzyme

7 Amount of enzyme – a greater amount of protein will be broken down per unit of time compared to the rate when protein is increased or vice versa.

8 Answers will vary.

1: Place 10 ml of the same concentration of protease in five test tubes

2: Prepare 5 × 1 cm³ pieces of the same protein

3: Leave one cube whole, cut one cube in half, cut one cube into quarters, one cube in into eighths; finely dice the last piece.

4: Label the test tubes: 1–5.

5: Place the uncut cube into test tube 1, after 5 minutes measure the amount of protein broken down.

6: Repeat with each of the four different protein samples.

7: Use a control: tube with no protein added.

4.3.4 Enzyme inhibitors p.100

1

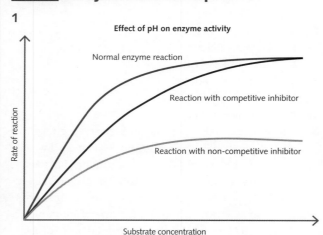

Effect of pH on enzyme activity

2 The normal enzyme reaction will show the fastest reaction rate as it is not being inhibited. Competitive inhibitors compete with the substrate for space in the active site, increasing the substrate concentration means that the substrate will enter the active site more frequently than the inhibitor, and the rate of reaction will increase. Non-competitive inhibitors do not bind to the active site, they do not compete with the substrate so adding more substrate will not decrease the inhibition, and the rate of reaction will not increase. Initially, when the non-competitive

inhibitor is added, it can bind to the substrate and it is often not released, slowing the initial rate of reaction as seen by the dip in the graph.

4.4 Chapter review p.102

4.4.1 Key terms p.102

A short-term molecule carrier of energy within the cell	Adenosine triphosphate
An unloaded coenzyme	NAD+
A loaded coenzyme	NADPH
Secondary binding site on the enzyme that a non-competitive inhibitor binds to	Allosteric site
Molecule with a similar chemical structure to the enzyme's substrate	Competitive inhibitor
Speed at which a biochemical reaction occurs	Reaction rate
One model of enzyme function	Lock and key model or induced fit
Place on the surface of an enzyme molecule where substrate molecules attach	Active site
Atoms are joined to make more complex molecules	Anabolic reaction
Variable that limits the rate of a reaction	Limiting factor
Sum of metabolic reactions in a cell	Cellular metabolism
Energy required to initiate a reaction	Activation energy
Enzyme that provides energy for the cell through synthesis of ATP	ATP synthase
Anabolic reaction using light energy to form glucose from water and carbon dioxide	Photosynthesis
Series of chemical reactions, each controlled by a specific enzyme, that converts a substrate molecule to a final product	Biochemical pathway

4.4.2 Exam practice p.103

1 C
2 A
3 A
4 C
5 D

Chapter 5 Biochemical pathways: photosynthesis and cellular respiration p.105

Remember p.105

1 Light
2 Chloroplasts
3 Inputs: light, carbon dioxide, water. Outputs: oxygen, (water), glucose
4 Cytosol and mitochondria
5 Inputs: glucose and oxygen. Outputs: carbon dioxide and water
6 ATP

5.1 Photosynthesis as a biochemical pathway p.105

5.1.1 **Structure and function of chloroplasts p.106**

1

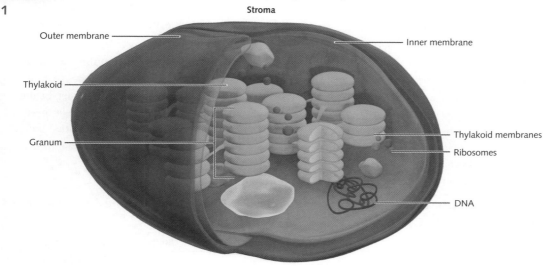

Stroma

Outer membrane

Inner membrane

Thylakoid

Granum

Thylakoid membranes

Ribosomes

DNA

2

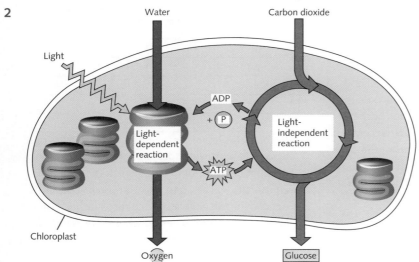

Water

Carbon dioxide

Light

ADP

+ P

Light-dependent reaction

Light-independent reaction

ATP

Chloroplast

Oxygen

Glucose

3 Carbon dioxide

5.1.2 **Photosynthesis in C₃, C₄ and CAM plants and the role of Rubisco p.107**

1 Carbon fixation, reduction, regeneration of RuBP
2 Stage 1, carbon fixation
3 As with other enzymes, Rubisco will denature at high temperature and low pH. The shape of its active site would be permanently altered. The substrates RuBP and CO₂ would no longer be able to enter the active site and the carbon fixation reaction would not occur. The Calvin cycle will be reduced or stop.
4 Rubisco's active site would be altered by the binding of the non-competitive inhibitor at an allosteric site separate from the active site. RuBp and CO₂ would no longer be able to enter the active site and the first step of the Calvin cycle, carbon fixation, could not

occur. The Calvin cycle would be reduced and not be not as efficient.
5 a No energy needed
  b Energy required: 6 ATP provides energy
  c Energy required: 3 ATP provides energy
6 They both go through the Calvin cycle but in C₃ plants carbon fixation is the first step of the Calvin cycle and this occurs in the mesophyll. C₄ plants have an additional step where carbon fixation occurs in the mesophyll cells producing a 4-carbon compound before the entering the Calvin cycle in bundle sheath cells which also contain chloroplasts. In C₃ plants, carbon dioxide enters the plants via open stomata which leads to water loss, in C₄ plants this does not occur so the stomata can be closed and water can be conserved.

7 a

C$_4$ plants	CAM plants
Carbon fixed during the day	Carbon fixed at night
Stomata open during the day	Stomata closed during the day
Takes place in bundle sheath cells	Takes place in mesophyll cells
Will have water loss	Reduces water loss

b C$_4$ and CAM plants are well adapted to their environments. CAM plants include cacti and succulents that survive the hot, dry climate of the desert by reducing water loss. C$_4$ plants will have water loss, but this is not as important in environments where water is found.

5.1.3 **Inputs and outputs of photosynthesis – Part A p.109**

1 Answers will vary. Set up two identical lamps attached to a retort stand in the same area to ensure they are at the same temperature. Position the lamps exactly 20 cm from the bench surface. In one 250 ml beaker place 200 ml of distilled water, place 200 ml of sodium hydrogen carbonate solution in the second beaker. Cut 20 leaf discs of similar size from a fresh leaf and remove air from the leaves' airspaces with a vacuum. Place 10 leaf discs in each beaker. Place one beaker under each of the lamps so the lamp is on the centre of the beakers. Turn on the lamps and time how long it takes for the leaf discs to rise in each beaker.

2 a Independent variable: presence or absence of sodium hydrogen carbonate

Dependent variable: time it takes for leaf discs to rise

b Temperature: ensure both beakers are in the same area; size of leaf discs: use identical hole punches to produce discs; type of plant: use leaves from the same plant; amount of light: measure distance from beaker to light, position under same part of light, lights should be same brand and voltage.

c The control is the beaker containing distilled water – to be able to compare differences when sodium hydrogen carbonate is present.

d All the leaves in the beaker containing sodium hydrogen carbonate will rise, no leaves in the beaker containing distilled water will rise.

e The leaf discs in both beakers will rise at the same rate or no leaf discs rise.

3 Answers will vary. Repeat the experiment a number of times; ensure all measurements are included in the method to control the variables; use appropriate measuring equipment; use consistent criteria to determine if leaves are floating or not.

4 Answers will vary. A well-designed method that includes all steps; ensure timing equipment is used correctly; measure distance from bench to lamps consistently; repeat the experiment a number of times, use consistent criteria to determine if leaves are floating or not.

5 Answers will vary. Burns from handling hot light globes: allow long enough time for globes to cool before handling; water and electricity: keep beakers away from power points, ensure leads on lamps are not damaged; sodium hydrogen carbonate is a mild irritant to eyes: wear safety glasses when making solutions.

6 Answers will vary.

Time taken for leaf discs to rise in distilled water and sodium carbonate solution

Time (mins)	Number of leaf discs floating	
	Distilled water	Sodium hydrogen carbonate solution
0		
1		
2		
3		
4		
5		

7 In the light-dependent reaction, energy from light is used to split water producing oxygen as a by-product, the production of oxygen in this stage causes the leaf discs to rise. Carbon dioxide is needed for the next stage of photosynthesis. It is the light-independent reaction where it is used to form glucose via the Calvin cycle. The leaves rising suggest that in the absence of carbon dioxide obtained from the sodium hydrogen carbonate the light-dependent reaction occurred at a slower rate due to ADP/P$_i$ and NADP$^+$ not being fed back into the light-dependent stage.

8 Answers will vary. Leaf discs in a solution of sodium hydrogen carbonate produced more oxygen due to photosynthesis than leaf discs in distilled water – this did not support the hypothesis If carbon dioxide is a key requirement of photosynthesis, then photosynthesis will stop if carbon dioxide is no longer provided, as photosynthesis appeared to continue but at a slower rate.

Limitations: We cannot be sure if it was photosynthesis causing the leaves to rise or another factor; errors in the method could have contributed to the results; for example, not removing air from airspaces correctly or the experiment was only carried out once.

5.1.4 Factors that affect the rate of photosynthesis p.112

Light p.112

1 Answers will vary. If different coloured light does effect the rate of photosynthesis then the number of bubbles observed will be different for each colour of light.

2 Light colour or wavelength

3 The number of oxygen bubbles produced

4 Sunlight: 26.8, Red light: 38.3, Blue light: 30, Green light:15.3

5 Red light

6 Green light

7 Sunlight is white light (all the colours of light mixed together). Exposure to sunlight gives a baseline to compare the exposure to different coloured lights.

8 Chlorophyll is a green pigment, it does not absorb green light (it is reflected), it absorbs violet, blue and red light. If only green light is provided no photosynthesis will occur as chlorophyll is unable to absorb energy at this wavelength, chlorophyll will be able to absorb all the energy provided in red and blue light leading to a greater rate of photosynthesis. The results showed that green light produced 15.3 bubbles on average, sunlight produced 26.8, blue light produced 30, and red light produced the most bubbles with 38.3.

9 The results show that the most amount of photosynthesis in *Elodea* occurs under red light. To increase production of photosynthesis, expose the crop to only red light, which will increase the photosynthetic rate and provide more chemical energy for the plants.

Temperature and carbon dioxide p.114

1 Answers will vary. If concentration of CO_2 and temperature affects the rate of photosynthesis then increasing the concentration of CO_2 and temperature will increase the rate of photosynthesis.

2 Concentration of CO_2 and temperature

3 Rate of photosynthesis

4 Carbon dioxide is an input for photosynthesis. An increased concentration of CO_2 will lead to an increase in the light-independent stage of photosynthesis and a greater production of glucose, until a point. Photosynthesis is an enzymatic reaction and the rate of reactions increase with increased temperature. The temperature must stay within an optimum range. If it is too hot, the enzymes in both stages of photosynthesis will denature and photosynthesis will be reduced or not occur.

5 Increasing light intensity will increase the rate of photosynthesis until a limiting factor occurs; for example, there is not enough CO_2, substrate or enzyme present to continue with the increased rate of photosynthesis. Temperature is also a limiting factor. As CO_2 and temperature are kept constant in this experiment, the rate of photosynthesis cannot increase further.

6 Answers will vary depending on the hypothesis. The results support the hypothesis. The plants exposed to 0.4% CO_2 and 25°C had the highest rate of photosynthesis. Plants exposed to 0.1% CO_2 and 15°C had the lowest rate of photosynthesis.

7 Answers will vary. Increased climate change could lead to higher temperatures and higher levels of CO_2 will lead to higher rates of photosynthesis, potentially higher crop yields or increased temperatures could denature enzymes needed for photosynthesis and plants will die.

8 Answers will vary. The results support the hypothesis: when plants exposed to a higher concentration of CO_2 and a higher temperature, they will have a higher rate of photosynthesis. The plants exposed to 0.4% CO_2 and 25°C had the highest rate of photosynthesis. Plants exposed to 0.1% CO_2 and 15°C had the lowest rate of photosynthesis. Further investigations: testing plants with higher temperatures, using different wavelengths of light, using different CO_2 concentrations.

5.1.5 Inputs and outputs of photosynthesis – Part B p.115

1 Answers will vary. If plants are closer to a light source then photosynthesis will occur at a faster rate, taking carbon dioxide out of water, which will become more basic.

2 Qualitative

3 Not very valid or precise

4 To act as controls to compare changes in the pH of water when no plant is present in tubes 5 and 6 to changes with the presence of a plant; and to see the effect of no light affected tube 6.

5 No, there was no colour change observed for both 10 cm and 20 cm distance from the light and the same results were shown for the test tubes without a plant (the controls).

6 Refute the hypothesis, the results show there is no difference in pH (CO_2 concentration) observed in plants at different distances from a light source, this indicates that the rate of carbon dioxide use and production is the same in plants at different distances.

7 Answers will vary. The hypothesis that plants closer to a light source would photosynthesise at a faster rate using more carbon dioxide and making the water more basic was not supported. The results show that plants at 10 cm and 20 cm distance from a light source had no change in the pH of surrounding water, indicating no change in carbon dioxide concentration.

Possible limitations: not possible to detect a colour change in the pH by eye, not a large enough difference in distance from the light, cannot be sure light from other sources did not affect the results; need to test a greater range of distances from the light; use a more sensitive method to measure carbon dioxide concentration (e.g. digital pH meter).

5.2 Cellular respiration as a biochemical pathway p.117

5.2.1 Cellular respiration p.117

1

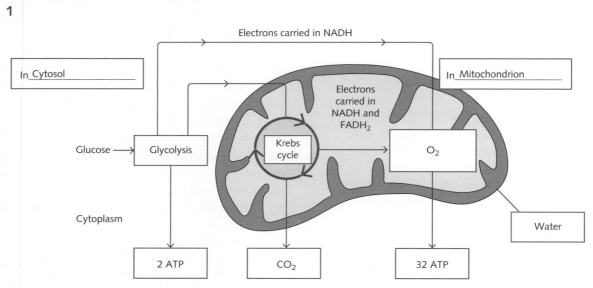

2 Answers will vary. The first stage of cellular respiration is glycolysis, this occurs in the cytosol. The input is glucose, which is broken down to two molecules of pyruvate, which releases energy used to make 2 ATP. Electrons and hydrogen ions that are accepted by NAD molecules become NADH. If oxygen is present the next stage occurs where the two pyruvates move into the matrix of the mitochondria; an intermediate reaction converts the two pyruvate into two molecules of Acetyl Co A, which enters the Krebs cycle. The outputs of the Krebs cycle are CO_2 (released as a waste gas), 2ATP and two types of loaded co enzymes – NADH and

$FADH_2$. The third and final stage of aerobic cellular respiration, the electron transport chain, occurs on the cristae, the folds in the inner membrane of the mitochondria. This stage uses the loaded coenzymes NADH and $FADH_2$ and oxygen. Electrons released from the loaded acceptor molecules pass along a chain of enzymes and compounds called cytochromes, releasing energy as they do so, this energy is used to make 32/34 ATP. Water is formed as a by-product when oxygen joins with the H^+ ions released by the loaded coenzymes. Cellular respiration produces 36/38 ATP molecules and H_2O and CO_2 as by-products.

5.2.2 Cellular respiration using oxygen p.118

1–4

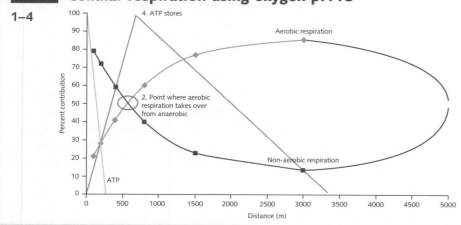

5 Anaerobic respiration is occurring in the cytosol of the cells, the process of glycolysis would be occurring in the cytosol of the cells. In glycolysis, the six-carbon molecule glucose is broken down with the help of enzymes into two molecules of pyruvate, a three-carbon compound. Two molecules of ATP are used to start the reaction and it produces four molecules of ATP per molecule of glucose, a net production of 2 ATP. During glycolysis two unloaded molecules of NAD+ are loaded with Hydrogen ions that are released when glucose is broken down to form two loaded NADH molecules. The molecules of pyruvate would then be broken down to form Lactic acid, in this process the two NADH are unloaded to become 2 NAD+ again, these 2 NAD+ can again be used in the glycolysis process.

6 Aerobic respiration is occurring in the cells. The first stage is glycolysis, which occurs in the cytosol, the six-carbon molecule glucose is broken down into two molecules of pyruvate, a three-carbon compound. Two molecules of ATP are used to start the reaction and it produces four molecules of ATP per molecule of glucose, a net production of 2 ATP. During glycolysis, two unloaded molecules of NAD+ are loaded with Hydrogen ions that are released when glucose is broken down to form two loaded NADH molecules. The pyruvate then enters the matrix of the mitochondria and is further broken down by reactions in the Krebs cycle to produce carbon dioxide, which is released as a waste product. The Krebs cycle produces 2 ATP and more loaded acceptor molecules: 6NADH and 2FADH$_2$ per glucose molecule. All the loaded acceptor molecules are then used in the last stage of aerobic respiration, the electron transport chain on the cristae of the mitochondria, the loaded acceptor molecules release their Hydrogen ions and electrons, the electrons are passed along a series of cytochromes, releasing energy which is used to produce 32 ATP. Oxygen is the final acceptor of the electrons and joins with the Hydrogen ions to produce water.

5.2.3 Cellular respiration without oxygen p.119

Anaerobic respiration
- 1 stage only: glycolysis
- All occurs in cytosol
- 2 ATP produced
- End product: pyruvate
- O$_2$ not required

Both
- Input glucose
- Glycolysis stage
- Production of ATP

Aerobic respiration
- 3 stages: glycolysis, Krebs cycle and electron transport chain
- Occurs in cytosol and mitochondria
- Makes 36/38 ATP
- O$_2$ required
- End products: CO$_2$ & H$_2$O

5.2.4 Putting photosynthesis and aerobic respiration together p.120

1 a M – light-dependent stage of photosynthesis

b N – light-independent stage of photosynthesis

c O – glycolysis

d P – aerobic cellular respiration

2 Water or inorganic phosphate

3 Reactions of the light independent stage of photosynthesis will no longer proceed

4

Removed molecule	Effect on		
	M	N	P
Oxygen	No change	No change	Decrease
Carbon dioxide	No change	Decrease	No change
Glucose	No change	No change	Decrease
Pyruvate	No change	No change	Decrease

5.3 Biotechnological applications of biochemical pathways p.121

5.3.1 Application of CRISPR-Cas9 technologies p.121

Gene editing with CRISPR p.121

1 Answers will vary.

Opinion:

Now, with CRISPR, new traits can be expressed and ready to go within two years.

CRISPR has the potential to make fruits and vegetables much healthier and more appetising. Gone will be the days of children hiding broccoli in their milk. Perhaps new varieties of fruit and vegetables can be created to join the trendy superfoods that are already on supermarket shelves.

'Certainly not, I don't want to start glowing in the dark', 'I might give it a go and see what it tastes like' 'If it's crisper, then it has got to be fresher right? So yes, I would try it.'

The research is controversial because many people do not like playing with nature. Some people believe that we do not have the scientific or regulatory ability to control unchecked evolutionary changes to species.

As the world's population grows, technologies such as CRISPR could come to be the saviour of our planet, allowing us to reduce the footprint of the agricultural industry and preserve wild ecosystems. CRISPR has the drawback of making it easier for people with bad intentions to do harm.

Anecdote:

The new gene-editing technology CRISPR is the talk of the scientific community. Scientists say that they will be able to genetically engineer foods much more easily. My sister has a friend who is a scientist, and she says that it could have potential impacts on my young children, so probably not.'

Fact:

It has the potential to be an exact, fast and economical way to edit DNA. This was a slow and tedious process that involved thousands of plants and hundreds of hours…(getting a gene) into a commercial crop and getting the product to market could take up to 10 years.Some people do not realise that CRISPR has been known about since 1987 as a bacterial defence mechanism. Since then, scientists have been busily unlocking its secrets. In 2012, CRISPR's potential for gene editing was discovered. Jointless tomatoes, fungus resistant bananas, non-browning mushrooms, and high yield corn, soy and wheat have been developed using CRISPR. Scientists have used CRISPR to systematically knock out one gene at a time and see what happens to the plant without that gene. This is how a group of researchers in Mexico are approaching the development of a non-browning avocado, but first they must locate the browning gene to knock it out. The researchers have collected the genomes of hundreds of varieties of avocadoes and are painstakingly sifting through the DNA to find differences in the genomes of browning and non-browning types.

2 Answers will vary. Does 'CRISPR modified' need to be on labels and displayed at point of sale? Do these products present any safety risks for consumers? How can the use of this technology be regulated? Who decides what characteristics should be modified? Social: greater nutrition, better health, acceptance of product. Economic: benefit supermarkets, less waste of product, greater yields for farmers. Legal: who owns specific modifications, privacy. Political: regulation of this technology.

3 Answers will vary. Students must provide evidence to support their opinion.

5.3.2 Anaerobic fermentation of biomass for biofuel p.122

1 The biomass comes from plants. Lignin and cellulose make up cell walls.

2 Carnivores do not eat plant material and therefore do not have lignocellulose as part of their diet.

3 It is possible to break down the materials in biomass by a catabolic or exergonic reaction, the breaking of the bonds releases chemical energy that can be converted into other forms of energy.

4 Glucose + 2ADP + 2Pi → ethanol + carbon dioxide + 2ATP

5.4 Chapter review p.124

5.4.1 Key terms p.124

1 1D, 2G, 3O, 4K, 5E, 6F, 7N, 8L, 9B, 10M, 11E, 12H, 13I, 14A, 15J

2 Answers will vary.

5.4.2 Exam practice p.126

1 D

2 C

3 C

4 D

5 B

6 A

7 a ADP or NADP+

b CO_2 or light intensity or temperature

c i To absorb different wavelengths of light. Chlorophyll-a absorbs light at the violet end and orange-red end, whereas chlorophyll-b absorbs best at blue and orange wavelengths.

ii These wavelengths are reflected, not absorbed, and accounts for why leaves appear green

d i At 10°C the activity of enzymes that catalyse the reactions of photosynthesis, have molecules with low kinetic energy; therefore, they will move around less and there is less chance that substrates will enter the active sites of enzymes and consequently photosynthetic reactions will occur more slowly.

ii 30°C is the optimal temperature for enzyme activity so the rate of photosynthetic reactions will be fastest provided there is enough light, water and CO_2, at low light intensity the reaction is limited by the amount of light available.

iii At 40°C in both plants, the bonds in enzymes have been altered, denaturing the enzymes needed for photosynthesis, the active sites of these enzymes have been permanently changed and the substrates are no longer able to fit in the active sites so the reactions will not occur, therefore little photosynthesis will occur.

Chapter 6 Responding to antigens p.129

Remember p.129

1 Any condition that interferes with the correct functioning of an organism, or a part of an organism

2 Answers will vary, include the disease and the part affected e.g. Tuberculosis affects the lungs.

3 Prokaryote, binary fission

4 Waxy cuticle: prevent infection (and water loss); hairs: prevent damage and infection caused by disease-carrying insects

5 Stomata by diffusion

6.1 Physical, chemical and microbiota barriers in animals – first line of defence p.129

1 Eyes (C): tears flush out foreign material and contain antimicrobial enzymes

2 Nose (P): mucus membranes in nose trap foreign material

3 Mouth (C): saliva contains antimicrobial enzymes

4 Armpit (C): sweat has a low pH, prevents pathogens from colonising

5 Urinogenital area (C): urine flushes away pathogens and microbiota, low pH prevents pathogens from colonising

6 Cut (leg) (P): blood clotting creates barrier to entry

7 Ear (C): wax contains antibacterial substances

8 Respiratory system (P & C): mucus traps pathogen, cilia moves pathogen to mouth and nose to be removed

9 Skin (P): intact/unbroken skin provides barrier to entry, secretions on skin such as sebum have antimicrobial properties

10 Stomach (C): stomach acid is low pH, kills bacteria and other pathogens

6.1.2 Microbiota as a barrier p.130

1 Table 6.1: 13, 23, 18, 21, 16, 0

2 AMC, FOX, GM, OX, P

3 AMC

4 Answers will vary.

Hypothesis: If household bleach and disinfectant solution has a greater inhibitory effect on the growth of *S.aereus* than ginger, garlic juice, eucalyptus oil and liquid soap then the zone of inhibition will be greatest around the household bleach and disinfectant solution.

I.V: type of substance used to control growth of *S.aereus*

D.V: the diameter of the zone of inhibition

Method: All equipment used must be sterile, wipe the bench and work area with bleach or ethanol before beginning.

Step 1: Set up a Bunsen burner, all steps should be performed close to the Bunsen burner to reduce chance of contamination.

Step 2: Prepare two sterile nutrient agar plates, label the base of the plates – divide into four quarters, label seven quarters near the edge of the plate with the names of the substances to be tested, label the last quarter control

Step 3: Prepare an identical concentration solution of each substance to be tested (the ginger should be juiced). Use a pipette to remove 1 ml of *S.aereus* culture, lift the lid off the labelled plate and transfer the bacteria to the surface of the agar.

Step 4: Spread the liquid evenly with the sterilised glass spreader, then replace the lid. Leave the plate on the bench for 2 minutes to allow the bacteria to penetrate the agar.

Step 5: Repeat steps 4 and 5 for the second plate.

Step 6: Sterilise forceps in the Bunsen burner flame, allow them to cool, then pick up a filter paper disc and carefully dip t into one of the seven solutions. Quickly touch the edge of the disc to the remains of the folded filter paper to blot, then gently place the disc on the correctly labelled quarter of the agar plate and replace the lid

Step 7: Repeat step 7 for the other six solutions

Step 8: For the last quarter repeat step 7 but dip the filter paper disc in distilled water, this disc will act as a control.

Step 9: Seal both agar plates and place in an incubator at 25°C for 24 hours

Step 10: Without opening the plates measure and record the diameter of the clear area around each disc (zone of inhibition).

Risk Assessment

Bleach, liquid soap and eucalyptus oil may irritate the eyes and nose.	Wear safety glasses, wash hands after handling
Eucalyptus oil is flammable and toxic	Keep well away from open flames and wash hands after handling

Results to support hypothesis: The zone of inhibition around the disinfectant and bleach discs will be larger than for the other substances.

Results table: (should have a title and column headings)

Zone of inhibition diameter for different household substances

Substance	Zone of inhibition diameter (mm)

6.2 Physical and chemical defences in plants p.133

6.2.1 Plants – first line of defence p.133

1

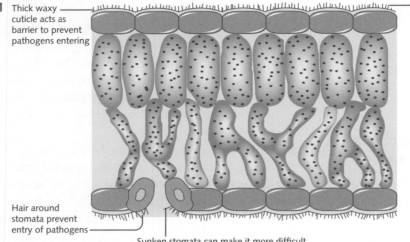

Thick waxy cuticle acts as barrier to prevent pathogens entering

Hair deter pathogens

Hair around stomata prevent entry of pathogens

Sunken stomata can make it more difficult for pathogens to enter the leaf

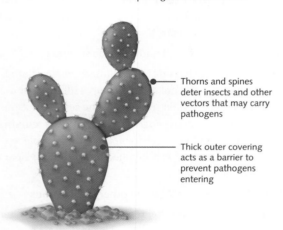

Thorns and spines deter insects and other vectors that may carry pathogens

Thick outer covering acts as a barrier to prevent pathogens entering

6.2.2 Plants – second line of defence p.133

1 Plants will not be able to carry out photosynthesis; without glucose cellular respiration cannot occur to produce ATP needed to provide energy for all cellular processes required to maintain life.

2 a Answers will vary. Can a chemical vaccine trigger a defensive chemical response in plants?

b Yes, the vaccine helped the plants switch on a defensive response that protected them from pathogens

3 Controlled experiment

4 A control to test that the response was not caused by a liquid being applied to the bottom of leaves but by the vaccine.

5 Answers will vary. Are treated plants safe for human consumption?

What is the best concentration of NHP to use to control black speck?

6 Controlled experiment to test any side effects of consumption using animals, correlational study (between treated and untreated plants) to analyse any differences in chemical content of fruit/vegetable/grain etc.

6.3 Innate response in animals – second line of defence p.135

6.3.1 Some specific cells of the innate immune response p.135

Suggested answer: answers will vary.

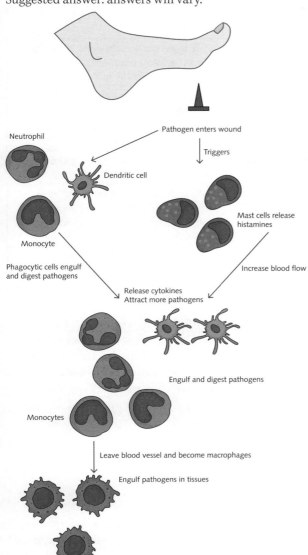

6.3.2 Innate immune response – inflammation p.139

1   Macrophages and dendritic cells, phagocytic white blood cells, identify the presence of a pathogen. They secrete cytokines that attract more phagocytic cells to the area.

Mast cells are granulated cells found in the tissues around the wound. They release their granules that contain histamines, which cause blood vessels to dilate and become more permeable.

Two types of phagocytic cells are neutrophils and monocytes; they are able to leave blood vessels and enter the tissues at the site of infection. They act as macrophages to engulf and destroy pathogens.

A clot will form to prevent further pathogens entering the wound.

All these events cause swelling, redness, heat and pain.

2   Inflammation is being resolved, cytokines are released to stop white blood cells being attracted to the site; blood vessels return to normal size and become less permeable. Macrophages are stimulated to clean up any cell and tissue debris at the site.

6.4 Antigens and pathogens p.140

1   Cellular: fungi, protists. Non-cellular: prions or virus

2

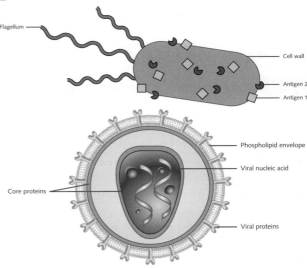

3   a   Bacteriophage/virus injecting its DNA into host bacterium.

b   Viral nucleic acid is transcribed and translated inside host bacterium to produce many copies of viral genome and protein coat.

c   Virus assembly begins. Viral genome is inserted in the viral protein coat.

d   The rest of the bacteriophage is assembled.

e   Bacterium bursts and releases the bacteriophages.

6.5 Allergens p.142

1   Redness, swelling and heat around the site of the sting, difficulty breathing, hives, itching

2   IgE antibodies are produced in response to the antigen and bind to receptors on mast cells. When the allergen binds to the IgE antibodies that are already attached to mast cells, the mast cells release histamine. Histamine causes inflammation in tissues.

3 a 0–4

b 65–69

c Blood types are tested and matched before transfusions; serum is purified before use.

d The number of unspecified causes of anaphylaxis increases from 0–30, increases more slowly from 30–60, where it stabilises and then decreases after 70 years of age.

e Answers will vary. After 80 years of age most people have determined their cause of allergies, aware of allergies to drugs, no new exposure to unknown allergens; people have died prior to 80 years.

f Reliable and valid. The data is a large sample group from all Australian hospitals; also, data comes from medical experts who are determining the cause of anaphylaxis.

6.6 Chapter review p.144

6.6.1 Key terms – spot the errors p.144

allergen	an antigen that is normally innocuous but, in some circumstances, can cause an overreaction from the immune system known as an allergy
antigen	a large molecule, usually a protein or polysaccharide, that generates an immune response
bacteriophage	a virus that infects bacteria
cellular pathogen	a disease-causing that is made up of one or more living cells such as bacteria or fungi
complement protein	a number of small proteins found in the blood that, when activated, promote chemotaxis, cell lysis and phagocytosis
defensins	small antimicrobial peptides secreted by virtually all plants and animals
dendritic cell	antigen presenting cell that phagocytose and present antigens to cells of the adaptive immune system
eosinophils	leukocytes that secrete powerful enzymes capable of rupturing multicellular organisms
histamine	a chemical released by mast cells and basophils that increases blood flow and the permeability of capillaries

inflammation	an innate response to infection or damage that causes pain, swelling, heat and redness
lysis	the process of a cell bursting
lysozyme	an antibacterial enzyme found in tears, saliva and other body fluids
macrophage	a large white blood cell that phagocytoses pathogens; originates as monocytes in circulation
monocyte	a white blood cell that circulates in the blood and matures into a macrophage when it moves from the blood into the tissues
neutrophil	a phagocytic leukocyte found in the blood and tissues
pathogen	an organism foreign to the body and is capable of causing disease
phagocytosis	process by which phagocytes engulf a particle or cell
prion	an infectious protein that can cause other unaffected prion proteins in the brain to take the affected form causing transmissible spongiform encephalopathies
vasodilation	widening of blood vessels, particularly arterioles

6.6.2 Exam marking p.146

1 a Correct according to marking guide, 1 mark.

b Gives physical barrier (hairs) but does not explain how this helps expel bacteria. 1 mark.

c Cell spelt correctly (eosinophil) (1 mark) and function given incorrect (eosinophils break down the cell wall not plasma membrane). 0 mark

Total marks: 3/5

2 a i 0 mark

ii 1 mark

b i 1 mark

ii 0 marks. Needed to say how a pathogenic agent is different to a pathogenic organism e.g. pathogenic agent can only be reproduced inside a living cell and a pathogenic organism can reproduce outside of a living cell.

Total: 2/4

Chapter 7 Acquiring immunity p.148

Remember p.148

1 The first line of defence is non-specific or also referred to as the innate immune response. It consists of chemical and physical barriers that prevent invaders from entering the body. The second line of defence involves cells, chemicals and processes that work to destroy invaders in a generalised non-specific way.

2 Pathogens are infectious disease-causing agents; antigens are molecules that trigger an immune response

3 The response is the same for all pathogens, it is a generalised response and not targeted to one specific pathogen.

4 PAMPs are pathogen-associated molecular patterns. These are found on the surface of pathogens. DAMPs are damage- or danger-associated molecular patterns that are released from damaged or dying cells.

5   PRRs are pattern recognition receptors on cells of the innate immune system (e.g. macrophages) that recognise PAMPs and DAMPs.

7.1 Adaptive immune response – third line of defence p.149

7.1.1 Lymphatic system – an analogy p.149

Table 7.1. Bank robbers – pathogens, bank walls – first line of defence e.g. intact skin, metal detectors – dendritic cells, security cameras – antigen presenting cells, security guards – helper T cells, Police officers – B cells and Tc cells, Wanted posters – memory B and T cells

7.1.2 Cells of the adaptive immune system p.150

1   Adaptive immunity: B lymphocytes (B cells) T lymphocytes (T cells), B plasma cell, Memory B cell, cytotoxic T cell (T_c cell), Helper T cell (Th cell), regulatory T cells (T_{reg} cell), memory T cell

In both adaptive and innate: dendritic cells and antigen presenting cells

7.1.3 Intracellular or extracellular pathogens p.150

1   A virus is an intracellular pathogen so antibiotics (which work against bacteria, an extracellular pathogen) would not be effective against a cold virus.

2   Answers may vary. Wear personal protective clothing: lab coat, face mask, gloves, safety glasses, use sterile of one-use disposable equipment, conduct all tests/processes in a sterile work area

3   Cytotoxic T cells or killer T cells

4   a   It is a virus or intracellular pathogen.

    b   T helper cells detect the antigen of the virus on an antigen presenting cell, the T_h cell releases cytokines that cause the activation and cloning of the T_c cells.

5   MHC class I markers display antigens (present on nearly all nucleated cells) from intracellular pathogens; MHC class II markers display antigens from extracellular pathogens (mainly found on cells of the immune system).

6   Antigen presenting cell (dendritic cell, macrophage or B cell) engulfs pathogen by phagocytosis → break down pathogens with the aid of lysosomes→ antigens from the digested pathogen are placed on a MHC class II marker→ the MHC class II marker with the

antigen is displayed on the outside of the antigen presenting cell.

7   Answers may vary.

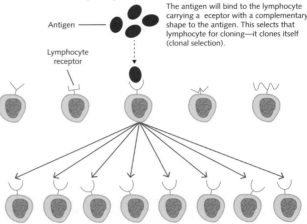

Clones of the lymphocyte that the antigen had bound to all carry identical receptors that are complementary to the antigen.

7.2 Humoral immunity p.152

1

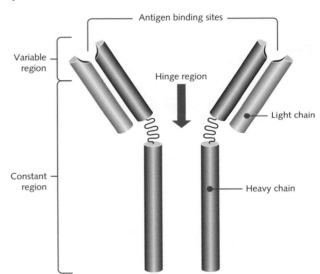

2   Antigen from the virus binds to a complementary antibody on the outside of a naïve B-cell. The B-cell becomes an antigen presenting cell and becomes activated, but the process is often further enhanced by a previously activated by a T helper cell. Once activated, the B-cell clones itself, some of the clones become plasma B-cells and some become B memory cells. The plasma B-cells produce antibodies that are complementary to the specific antigen.

7.3 Cell-mediated immunity p.153

7.3.1 Cell-mediated response p.153

1   The T helper cell recognises its specific antigen on an antigen presenting cell by binding to the antigen with a complementary receptor. The T_h cell releases cytokines. The cytokines activate cytotoxic T cells;

T_c cells inspect body cells and bind to their specific antigen if found on a MHC class I marker of a body cell. The T_c cell then releases cytotoxic proteins that destroy the infected body cell.

2   Humoral response involves B cells: B-cells are activated when they bind to antigens on extracellular pathogens, in response they clone themselves, and produce B memory cells and B plasma cells. The B plasma cells produce and release antibodies into the blood stream, these have two complementary binding sites for the specific antigens on COVID-19 virus. They can bind to the virus antigens which assists their elimination by phagocytic cells.

The cell-mediated response involves T cells: T cells are effective against intracellular pathogens and are activated when they bind to their specific antigen on the outside of an antigen-presenting cell. The effectors are cytotoxic T cells that use receptors to identify their specific antigen on MHC class I markers on body cells. If they identify their antigen, they release toxic chemicals that cause the infected cell (cell infected with COVID-19) to be destroyed.

7.3.2 Adaptive immune responses p.154

1   Similarities: both B- and T-cells carry regions (antibodies on B-cells and specific T-cell receptors) that recognise one specific antigen and attach to that antigen via a complementary binding site.

Differences: Humoral response: B-cells – specific antigen-binding site is located on antibodies; B cells recognise and bind to free antigens found outside cells

Cell mediated response: T-cells – specific antigen binding site on receptors, can recognises antigens attached to MHC markers on the outside of cells,

2   Yes. Cell mediated response – the T-cell to be cloned occurs when a particular Th cell finds its specific antigen on the MHC class II markers of an antigen presenting cell. The Th cell binds to the antigen with complementary receptors and releases cytokines, this causes that Th cell to clone itself many times

Humoral response: a B-cell encounters its specific antigen and binds to it via complementary antigen binding sites on its antibody The B-cell then becomes an antigen presenting cell, a Th cell with a receptor for the same antigen binds to the antigen presented by the B-cell. The B cell is now activated and produces a large clone of itself

3   Humoral: plasma cells, cell mediated: cytotoxic T-cells

4   Memory B cells, memory T_h cells, memory T_c cells. These cells remain in the body in greater numbers and provide a faster and stronger response if that specific antigen is encountered again.

Memory B-cell   Memory T_H cell   Memory T_C cell

7.4 Active and passive immunity p.156

1   a   Anti-venom contains antibodies that are complementary to the antigens in snake venom.

b   Active immunity – a person mounts their own immune response to an antigen and produces specific antibodies and memory cells specific for the antigen; this produces long-term immunity.

Passive immunity – a person is given antibodies from another source; they do not produce their own antibodies or specific memory cells; this does not provide long-term immunity.

c   i   No, this is artificial passive immunity as the antibodies have been delivered via an injection.

ii There would be no memory cells specific for the venom.

2   a   It takes several days for the adaptive immune response to occur and for a large number of specific antibodies to be produce on first exposure to the antigen; therefore, at day 12 there are a large number of specific antibodies present

b   Between day 12 and day 20, the antigen has been eliminated so production of antibodies reduces; antibodies have a short life, and number of antibodies decrease.

c   When the antigen is introduced a second time, the presence of memory cells from the first vaccination means that the antigen will be recognised more quickly. Therefore, between day 36 and day 40 there will be a more rapid and larger production of plasma cells and therefore antibodies specific to the antigen.

d   By day 50–60, the antigen has been eliminated and number of antibodies decreases but stays at a higher level compared to after first vaccination due to an increased number of B memory cells. It will not return to zero due to the presence of the memory cells.

7.5 Chapter review p.158

7.5.1 Key terms p.158

Student responses will vary.

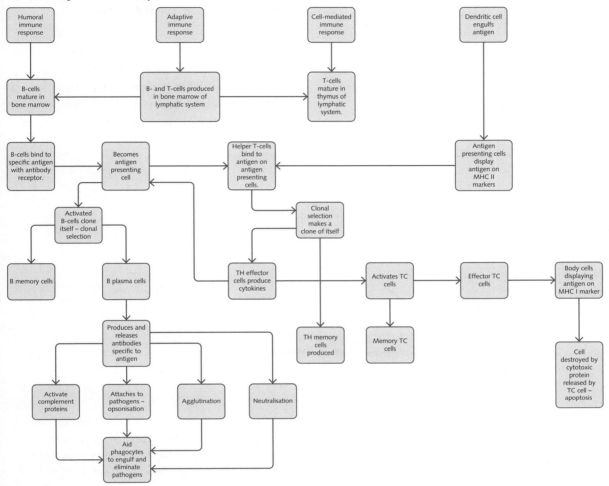

7.5.2 Exam practice p.159

1 D
2 D
3 B
4 a Strain N, P or Q
  b All of them.
  c Student X, would bind to the antigen on strain P as it has a complementary binding site, students Y and Z are not correct antibodies as both antigen-binding sites must be identical.
5 a Z – This cell produces the least percentage of plasma cells, so will produce the smallest number of specific antibodies needed to eliminate the pathogen.
  b Plasma cells: produce specific antibodies that can bind to the antigens on a pathogen that help clump together pathogens and assist phagocytic cells in eliminating the pathogen by phagocytosis.
  Memory cells: allow the pathogen to be recognised more quickly and cause a more rapid and larger production of plasma cells and specific antibodies against the pathogen so it is eliminated before it can cause symptoms.

Chapter 8 Disease challenges and strategies p.162

Remember p.162

1 A disease is any condition that affects the structure or normal functioning of an organism.

2 An infectious disease is caused by a pathogen and can be spread to other members of a population.

Non-infectious diseases are not caused by pathogens, the cause can as a result from genetic or environmental reasons, and cannot spread to other members of a population

3 Disease can be caused by pathogens, genetics, environmental factors, or lifestyle

4 a Molecule, often a protein, that causes an immune response

b 'Y-shaped' protein produced by B-plasma cells in response to antigens that has two specific antigen binding sites and can bind to antigens.

c Signalling molecules of immune cells that are used to coordinate immune responses

2

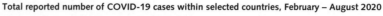

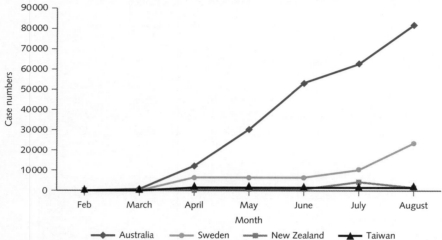

Total reported number of COVID-19 cases within selected countries, February – August 2020

Case numbers (y-axis) vs Month (x-axis: Feb, March, April, May, June, July, August)

Legend: Australia, Sweden, New Zealand, Taiwan

3 *Australia*: The numbers of report cases of COVID-19 increased in first month, from March to April then the numbers stabilised and increased again during August.

New Zealand: Only a slight increase to approx.1400 cases in April and then stable numbers of cases persisted until August

Taiwan: A slow increase to around 400 cases that persisted until August

Sweden: A steady and rapid increase of COVID-19 cases from Feb to August

4 Answers will vary.

New Zealand was successful at controlling the outbreak: implementing quick and strict lockdown, isolating people from each other (which successfully stopped community transmission), 14-day quarantine for people entering country (stopped introduction of the virus), high testing (enable quick identification of those infected to stop the spread).

8.1 Emerging and re-emerging pathogens p.163

8.1.1 Spread of pathogens in a globally connected world p.163

1 Modern transport allows rapid movement of people between populations within and between countries; people can move between different populations in less time than it takes symptoms of infection to occur. (Incubation periods are often longer than 24 hours and can take up to two weeks.)

Taiwan was also successful at controlling the outbreak: limited spread in population by strict monitoring of people entering country (stopped entry of virus into the population), quarantined infected people (prevented further spread) and implemented face masks (limited the infection via airborne virus).

Australia relatively successful: social distancing guidelines, working from home, closing schools (limited the spread between individuals), border closures and quarantine (stopped new cases from entering the country and limited the spread), large testing regime (identified cases for quarantine) and contacts tracing (limited the spread).

Sweden was not successful in limiting the spread, herd immunity is only effective when a large proportion of the population is immune, and this did not occur; many people needed to be infected and recovered to allow this. No restrictions on social gathering or movement allowed the virus to continue to spread between people.

9780170452618

5

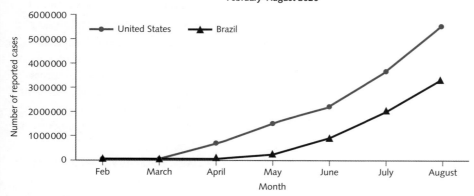

Total reported number of COVID-19 cases within the United States and Brazil, February–August 2020

6 Positive exponential increase in cases of COVID-19 cases over time in each country, Brazil's graph is roughly two months behind the United States.

7 Answers will vary. No nationwide rules about staying safe regarding masks, social distancing; no hard lockdown; focus on economic damage rather than human damage; and this all leads to large numbers of infections and deaths compared to countries that went hard early on.

8 Answers will vary. Close borders quickly, implement strict quarantine for people entering the country, extensive education program for the population, put in measures to prevent the spread (e.g. face masks, social distancing), have an effective and rapid contact-tracing system.

8.1.2 Impact of European arrival on Aboriginal and Torres Strait Islander peoples, p.167

Answers will vary. These points are given as examples, students should not just summarise.

» The first Europeans to arrive in Australia in the 18th century also brought many pathogens with them.

» While these infectious diseases were common in European populations, the indigenous population of Australian had never been exposed to them before.

» The indigenous population had no immunity to these pathogens, so they were reliant on their innate immune response to protect them on first exposure to these pathogens.

» Many ATSI peoples died, up to 50% in the Port Philip area, from the diseases before their adaptive immune response could eliminate the pathogen.

» The surviving population were prevented by the government and settlers from continuing their traditional hunter-gatherer lifestyle and access to some of their traditional medicine plants was restricted. The government had to provide

rations to allow the indigenous population to feed themselves. This lack of access to their normal varied diet and their traditional medicines would also have had a detrimental effect on the health of the ATSI peoples.

8.2 Strategies for controlling pathogen transmission p.170

The virus can be spread in droplets expelled from the mouth and nose. So:

» coughing into arm limits the release of droplets into the air that can carry and spread the virus

» using a tissue also prevents the release of droplets containing the virus into the air, prevents spread of the virus

» disposing of the tissue protects other people from coming into contact with the virus on the tissue; as an infected person's tissue will now carry virus

» washing your hands with soap and water eliminates the virus and stops infection when you touch your mouth or nose since the virus can survive on surfaces; you can pick up the virus on your hands when you touch contaminated surfaces.

8.3 Vaccination programs p.171

8.3.1 Herd immunity p.171

1 Measles has higher reproduction number, 15, than the common cold, approx. 2.

Measles also has a higher fatality rate, 0.2–0.3, compared to the common cold, 0 fatality rate.

This represents the number of people an infected person can transmit the pathogen to. Therefore, a greater % of people infected with measles would die if an outbreak occurred as it is more contagious than the common cold, so hence, it would be more deadly.

2 Contagiousness is a measure of how many people an infected person can spread the pathogen to. Infectious diseases with a high fatality rate tend to be less contagious. Figure 8.2 shows that diseases with a high fatality rate have a lower contagiousness, less than 5, with a few exceptions.

3 Ebola is less contagious than measles, each person infected with measles will spread it to 15 people, while people with Ebola will only infect approximately two people.

4 a For herd immunity to be effective, a large percentage of the population must have immunity; this prevents the pathogen from spreading as there are fewer hosts it can infect.

   If only 34 % of the population are vaccinated this will not provide herd immunity, roughly 7 in 10 people can be still be infected if they come into contact with a person carrying the virus, allowing the virus to spread between hosts.

   b To reduce the number of people in contact with each other, social distancing to reduce spread of the virus as it is spread by contact or breathing in airborne droplets released by infected individuals could be implemented.

   c Face masks, education campaign, promoting good hand hygiene, testing and contact tracing, quarantining infected people.

8.3.2 Vaccines p.173

1 a The immune system is not fully developed until this age; newborns who are breastfed are protected via natural passive immunity – antibodies received through breast milk will protect them from pathogens. It takes time for the baby's immune system to be able to produce its own antibodies.

   b Artificial immunity – the microbe, or parts of it, is injected into the person so that the immune system is stimulated (before they are exposed to it naturally).

   c Active immunity – the vaccine triggers an immune response and the individual produces their own memory cells and antibodies specific to the antigen.

   d The first vaccination will create memory cells and specific antibodies, each subsequent vaccination will create a larger response due to the pre-existing memory cells and a greater number of memory cells and specific antibodies will be created. This is necessary to ensure there are enough memory cells to provide immunity.

2 a Vaccination programs create herd immunity, if a large % of the population is vaccinated then this protects individuals that are not vaccinated as it is not possible for the pathogen to spread between hosts in the population. Over time the % of vaccinations in populations has increased, increasing herd immunity, and reducing deaths among the unvaccinated.

   b Answers will vary. Students should use data from the graphs. Vaccinations have greatly reduced deaths: before the introduction of vaccines diphtheria killed 4073 per decade, whooping cough nearly 3000, tetanus 625, polio and measles over 1000 each. Vaccinations have reduced this down to 0 or less than 10. Some of these diseases still exist in low numbers in the population and if children are not vaccinated, infection rates would again increase leading to deaths.

3 a Should healthy people be infected with COVID-19 to allow a vaccine to be tested?

   b Consequence-based approach

   c Answers will vary. Do the benefits outweigh the risk to the participants? What measures will be put in place to reduce the risk to participants? Justice: if participants are being drawn from communities with high incidence of infections, will this mean that people with socio economic disadvantage are used as guinea pigs to benefit others? Respect: informed consent, will participants be provided with enough information in an understandable format?

8.4 Immunotherapy strategies p.175

8.4.1 Monoclonal antibodies and the treatment of cancer p.175

1 Fuse a plasma B cell that produces a specific antibody with a cancer/tumour cell

2 It can divide repeatedly (immortal), produce identical antibodies.

3 They are produced by clones of the same hybrid cell and are therefore identical.

4

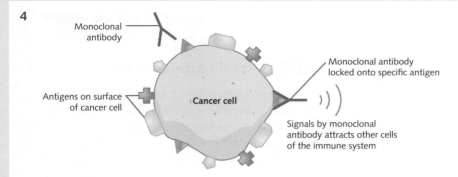

Monoclonal antibody

Monoclonal antibody locked onto specific antigen

Antigens on surface of cancer cell

Cancer cell

Signals by monoclonal antibody attracts other cells of the immune system

5 They can deliver targeted treatment to cancer cells that carry the specific antigen, less side effects, more effective at delivering treatment to the cancer cells.

8.4.2 Cancer and autoimmune diseases p.177

1 To investigate the protocols that should be put in place before clinical trials of new drugs and the biological molecules used in treatments before they progress to human trials to ensure safety of patients

2 Case study

3 Answers will vary. Carefully check medical history; avoid bias when selecting volunteers, use a wider demographic of volunteers e.g. gender, age.

4 Volunteers receive treatment intravenously and then observed.

5 Answers will vary. The drug trial did not deliver expected results; the drug had an adverse effect on healthy patients.

6 Improving informed consent of participants, clinical follow up of patients that experience adverse effects to fully investigate causes, better communication between clinical investigators and trial subjects.

7 Referencing should follow format: Author(s) name and initials, 'Title of article', *Title of journal*, available publication information (volume number, issue number, accessed date dd/mm/yyyy, <URL address>

Morris, A 2004, 'Is this racism? Representations of South Africa in the Sydney Morning Herald since the inauguration of Thabo Mbeki as president'. **Australian Humanities Review,** no. 33, accessed 11 May 2007, <http://www.australianhumanitiesreview.org/archive/Issue-August-2004/morris.html>

8.5 Chapter review p.179

8.5.1 Exam practice p.179

1 A
2 C
3 A
4 B
5 D

6 a Social factors: increased population coming into contact with wild animal habitat, urbanisation of habitats in proximity to wild animals increases the chance of direct transfer, transportation of wild animals, lack of education about transmission of disease.

(Note: Transmission is not through food or water.)

b Identifying viral pathogen allows identification of mode of transmission – processes can be put in place to prevent transmission, can determine correct antiviral drug to use to treat the pathogen, incorrect identification can lead to continued spread OR Identifying the virus allows the identification of antigens allowing the development of a vaccine, if the population is vaccinated the spread of the disease will be stopped.

c Answers will vary. Inspect and fumigate cargo entering new countries, quarantine people who come from countries in which the disease occurs, control mosquito breeding places (use mosquito nets or fly screens on windows), education campaign about suitable clothing/use of insect repellent.

Chapter 9 Genetic changes in populations over time p.181

Remember p.181

1 a a thread-like structure made of nucleic acids and proteins that encode genetic information

b a linear polymer built from amino acid monomers

c small, circular DNA independent of the chromosome in prokaryotic cells

d an infectious agent capable of causing disease

2 a Genotype is an organism's genetic composition, the expression of these genes in the organism produces the phenotype.

b Homozygous – two of the same alleles present for the gene locus, heterozygous – two different alleles for the gene locus

c A gene is a sequence of DNA that codes for a protein, genes have alternative forms called alleles.

3 The linking together of complementary nitrogen base pairs by hydrogen bonding. In DNA, adenine pairs with thymine and cytosine pairs with guanine.

4 The process by which the information in a gene is turned into a polypeptide, by transcription and translation.

5 Transcription provides a mobile copy of the DNA code in the form of mRNA that can carry the code out of the nucleus to the ribosomes. Translation is when the ribosomes read the mRNA code to create a sequence of amino acids that forms a polypeptide chain and the primary protein structure.

6 A vaccine delivers an attenuated form of an antigen from a pathogen to the body to trigger an adaptive immune response that leads to the production of memory B- and T-cells via the humoral and cell mediated immune responses. The memory cells are long lasting and specific against the original antigen that the body was exposed to.

7 Herd immunity is when a large proportion of a population is immune to a particular pathogen. This limits the ability for the pathogen to pass between hosts and protects individuals that are unable to be vaccinated and who are not immune.

9.1 Mutations: the source of new alleles p.182

9.1.1 Point mutations p.182

1 Mitosis produces somatic (body) cells. A mutation that occurs as the result of a mistake in mitosis will only affect the cell in which it occurred and its daughter cells (the somatic cells). Meiosis produces gametes or sex cells. A mutation due to mistakes in meiosis will be present in a sperm or egg cell, this is called a germline mutation. This will be present in all cells of the organism created with this mutated sperm or egg cell and can be passed to this organisms' offspring.

2 b Substitution – silent mutation, has no effect on translation as amino acid is the same

c Substitution – missense mutation, will produce a different amino acid sequence, the third amino acid will be Lys instead of Glu

d Insertion – frameshift mutation, all the triplet codes downstream from the insertion will change, translation will produce a different amino acid sequence in the polypeptide

e Deletion – frameshift mutation, all the triplet codes downstream from the deletion will change, translation will produce a different amino acid sequence in the polypeptide

9.1.2 Cystic fibrosis mutation p.184

1 Ile–Ile–Phe–Gly–Val (Work out template strand first, then the mRNA and then amino acids.)

2 a CTT
b Ile–Ile–Gly–Val
c Phe

3 The deletion of the three bases CTT alters how the DNA triplet code is read from the point of the deletion on, the deletion of the C from the second triplet and the TT from the third triplet, shifts how all the triplet codes are read downstream from the second triplet on.

9.2 Chromosome rearrangements p.185

Activity: illustrate block mutations p.185

1 **a** Deletion
Suggested answer

b Inversion
Suggested answer

c Translocation
Suggested answer

d Duplication
Suggested answer

Bacterial conjugation p.186

1 **a**

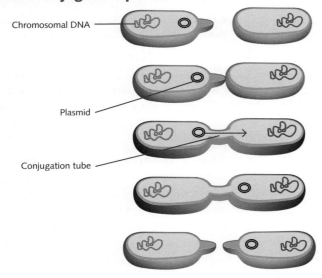

Chromosomal DNA

Plasmid

Conjugation tube

b Horizontal gene transfer is when the DNA from one bacterium is passed to another bacterium and becomes part of its genome, for example by conjugation. Vertical gene transfer in bacteria is passing on genetic material by reproduction, from parent to offspring by binary fission.

c It would introduce new genes to a bacteria population/colony and potentially advantageous traits that could spread through the population very quickly. It would speed up the rate of evolution.

9.3 Changing allele frequencies in populations p.189

9.3.1 Fowler's toad and the American toad p.189

1 Location 3

2 There is gene flow at locations 5 & 6, then interbreeding at the margins with location 3. Individuals from location 3 interbreed with those from location 7 who then interbreed with location 2. Location 2 overlaps with location 1 so there is gene flow.

3 **a** The Fowler's toads in Location 1 could be geographically isolated from the other populations making interbreeding with other populations difficult or it could be a smaller population.

b Answers will vary. Field work to observe the toads from Location 1 to see if toads from other locations were able to reach and interbreed with them. Could also use PCR analysis of DNA to see if any alleles from location 1 occurs in location 6.

4 The populations would no longer be able to interbreed, there would be no gene flow between the two populations, mutations that arise in one population will not be introduced to the other population. This could eventually lead to the gene pools in each of the populations being different and new species forming as different selection pressures could occur on each population.

9.3.2 Bottleneck effect p.190

1 Bottleneck reduces a population to a very small number due to catastrophic event (e.g. natural disaster), which kills organisms randomly, irrespective of phenotype, and leads to lowered genetic diversity. For example, koala population sizes were dramatically reduced by bushfire, the characteristics of the koalas did not determine if they lived or died.

2 Bushfires lead to bottlenecks as they kill many organisms indiscriminately and have the potential to significantly reduce the size of populations.

The bushfires of 2019/2020 killed thousands of koalas, dramatically reducing the genetic variation of the population.

3 Founder effect is when a new population is established by a few individuals in a different location to the parent population. The founder population's gene pool is less diverse than the parent population as it will only contain the alleles carried by the original founders. Example: the koala population of French Island and Kangaroo Island that were created with only a few founder koalas from a mainland population.

4 **a** 10-Pine Creek NSW and 15-French Island Vic

b Low genetic diversity: when compared to other populations, there are a number of populations in which there is very little genetic difference between the populations or where there is very high genetic difference. This suggests a lack of diversity in the Kangaroo Island population, either they are all alike another population or they are all different to another population.

c 14-East Gippsland, there is the least genetic difference between this population and the Kangaroo Island population (0.01), suggesting they share the same mutations in their gene pool.

d Gene flow could introduce new alleles into the gene pool of the Kangaroo Island koala population, creating greater genetic diversity, creating more variation in phenotypes and increasing the population's chance of survival.

e Answers will vary. Introducing new koalas into populations with low genetic diversity to introduce new alleles into the gene pools; captive breeding programs to breed koalas with desired genetic variation and release into wild populations; consolidating different populations to create a greater gene pool and more diversity.

9.4 Natural selection p.193

9.4.1 Selection pressures p.193

Figure 9.8 – no snakes circled

Figure 9.9 – three snakes circled

1 Predator – the mountain fox

2 A germline mutation occurred during meiosis

3 The white snakes

4 The frequency of the recessive allele will decrease but may not disappear as it can still be carried by heterozygous black snakes.

5 a Street lights now provide more light at night

b The black snakes

c The recessive white allele will increase or could become the only allele for skin colour in the population. The white allele will become fixed, as all the black snakes will be eaten removing the allele for the dominant trait of black skin.

9.4.2 Principles of natural selection p.194

1 a For natural selection to occur there must be differences in the phenotypes of individuals in a population. In the lizard population, there was existing variation in leg and arm length, and size of toe pads.

b The variations in the size of the lizards' arms, legs and toe pads could have been the result of a germline mutation, allowing this variation to be passed on to offspring.

c If all offspring produced by the lizards survived the population would increase. There are limited resources on the island, and it would not be able to support an increase in lizard numbers. The lizards must compete for the available resources and those that are better adapted will survive.

d The hurricanes on the island gave the lizards with shorter legs a selective advantage in that they were able to stay connected to branches, while the lizards with longer legs were blown off and killed. The lizards with short legs were able to survive in greater numbers, reproduce and pass on their alleles for short legs to their offspring.

9.4.3 Extinction p.195

1 Species extinction occurs when all members of a species have died, there are no living members of the species left. Low genetic diversity makes a species more vulnerable to extinction. Once a species is extinct all the genes of that species are lost.

2 The possible extinction of species not only removes the species from the planet but also our access to proteins produced by the genes of the species that may be useful as drugs.

3 a Answers will vary. Social – effect on the lifestyle of indigenous people; removal of the source of possible drugs to treat health issues; improve the income of farmers; more food produced for population; effect on global warming. Economic – financial gains for farmers and the country from increased agriculture; greater supply of crops may lower food prices; climate change impacts on economy. Political – who owns the land; how should land clearing be

controlled and regulated; voting backlash against government in power. Legal – who owns the land? compensation for indigenous people?

b Answers will vary. Social – potential to improve life quality of many people, possibility of cheaper and better drugs and therapeutics. Economic – financial gains of pharmaceutical companies, lower health costs. Political – regulating/legislating environmental protection laws, legislating who owns the drugs developed. Legal – who owns the drugs? Intellectual property issues

9.5 Human manipulation of gene pools p.196

9.5.1 Artificial selection: animal and plant breeding p.196

1 Answers will vary: Social – cheaper due to greater yield and prolonged shelf life, affordable to more of the population. Economic – greater yield more income for farmers, less loss of crop to disease. Legal – should someone be able to own the rights to selectively bred banana varieties? Political – how is selective breeding legislated? Bananas are now sterile, should the governments have prevented this?

2 c Step 1: The breeders would have evaluated the variation present in the bulldogs' skulls due to heritable mutations.

Step 2: Favourable allele – breeder selected the phenotype that was most advantageous to the needs of the breeder, this would not necessarily be a trait that would have given the bulldogs a survival advantage.

Step 3: Passing on favourable alleles. The breeders only bred the dogs that had the traits they had selected.

Step 4: Reproduce – these dogs would pass on the alleles for the selected traits to their offspring.

3 Answers will vary. Bioethical approach - virtues-based - that the breeder will act in an ethical way. Ethical concepts - non-maleficence - some degree of harm is justified by the benefits for the breeder.

Social factors - people can have access to the features in a dog that they desire; a dog built to purpose.

Economic - the breeder can sell the dog for an inflated price; dog may suffer from inherited disease which will cost more money for veterinarian fees.

9.6 Natural selection and consequences for disease p.198

9.6.1 Natural selection explains antibiotic resistance p.198

1 a Countries that show higher use of antibiotics also have a higher clinical resistance in *E. coli*. There are a few exceptions: Portugal and Italy show medium use, but high resistance and Greece has high use but medium resistance. The general correlation shown is that greater use of antibiotics produces greater clinical resistance in *E. coli*.

b Answers will vary. Whether the population follows directions when using antibiotics; types of antibiotics used by these countries; what antibiotics are prescribed for in these countries; introduction of resistant strains to Portugal unrelated to antibiotic use; how data of antibiotic use was collected.

2 Students need to include a key to indicate which bacteria are resistant or susceptible. The number of resistant bacteria increases in each circle.

Bacteria show natural variation in terms of resistance to antibiotics with some being resistant and some being susceptible. This variation is heritable and can be passed to the offspring. Antibiotics act as a selection pressure; they will kill the susceptible bacteria. The resistant bacteria have an advantage and are able to survive. The surviving resistant bacteria reproduce and pass their alleles for resistance onto their offspring. Over time the number of resistant bacteria in the population will increase.

9.6.2 Viral antigenic drift and shift p.200

1 Diagram a shows antigenic drift. Mutations in the viral nucleic acid lead to changes in viral antigens (proteins). If the antigens change enough a new strain will be created. Antigenic drift means the virus will no longer be recognised by the adaptive immune response if it reinfects the host. Diagram b shows antigenic shift. When two different influenza viruses infect the same host the genetic material from the two viruses can be shuffled and combined to create a new virus subtype. This virus will no longer be recognised by the adaptive immune response as the original virus. Antigenic shift and drift limit the ability to develop immunity to influenza virus, as each viral infection can be a slightly different version with different antigens not recognised by memory cells of the adaptive immune response.

2 a Antigenic drift: mutations in the human influenza virus

b Antigenic shift: new virus strain produced by shuffling genetic material of an animal and human virus

9.7 Chapter review p.201

9.7.1 Key terms p.201

Number	Letter
1	K
2	G
3	B
4	N
5	A
6	I
7	M
8	O
9	C
10	E
11	L
12	H
13	F
14	D
15	J

9.7.2 Exam practice p.202

1 a Mutations can create new alleles that produce new phenotypes (1), sexual reproduction (meiosis) creates different allele combinations due to recombination (1), random joining of gametes during sexual reproduction creates different genotypes in individuals and different phenotypes (1)

b i The populations on the smaller islands are a result of the founder effect (1). The gene pools of the small island populations will show less variation as all the alleles in the gene pool came from the two introduced founding lizards (1).

ii A selection pressure on the small island gave the lizards with larger sticky toepads an advantage and they survived in greater number, had a greater rate of reproduction and passed the alleles to their offspring compared with the lizards that have the smaller sticy toepads (1). Over generations the number of lizards with large sticky toepads in the population increased. (1)

2 a Answers will vary: the extinction event could have been due to: hunting, clearing of habitat or predation by foxes (1).

b **i** All the alleles present in the released eastern barred bandicoots

   ii A diverse gene pool is the most advantageous as more alleles means greater variation in the phenotypes of the bandicoots (1). This will improve the chance that there will be a phenotype that can survive if their environment changes of a new selection pressure is introduced (1).

c Founder effect: a new population is established by a few organisms; the gene pool will be less diverse than the parent population. The allele frequency of a recessive allele can be more prominent than in the parent population.

Bottleneck effect: a catastrophic event occurs that reduces a population down to a very small number, when the population recovers it will have less genetic diversity than the original population. The frequencies of alleles will change in the population based on chance events.

d Any one of: monitor numbers, monitor health, monitor breeding, monitor the presence of predators

Chapter 10 Changes in species over time p.204

Remember p.204

1 Mutations

2 Proteins are produced at a ribosome by translation, the order of the amino acid in a protein is determined by the bases in the mRNA with one amino acid being coded for by three bases. The sequence of amino acids in the protein can change due to a single base being deleted or added or swapped for another in the creation of the mRNA during transcription.

3 Variation is created by heritable mutations, variation is inherited, there is selection pressures in the environment and those individuals that are better adapted will survive and reproduce, successful variation accumulates over time within a population.

4 Low genetic diversity, human activities, environmental change

10.1 Studying fossils p.205

10.1.1 Fossilisation process p.205

Answers will vary. The summary should include the fish dying, rapid burial by sediments, avoids decomposition, lack of oxygen, sediment layers building up on the fossil, formation of sedimentary rock, explanation of mineralisation, left undisturbed and free from predators.

10.1.2 Relative dating techniques p.206

1 Area 1 – layer E. The deepest layer in area 2 E matches with layer C in area 1, so it is younger than layer E in area 1. The deepest layer in area 3 matches with layer D in area 1 so layer E in area 3 is younger than layer E in area 1.

2 1E (Oldest), 1D & 3E, 1C & 2E & 3D, 2D & 3C, 2C, 2B (Youngest)

3

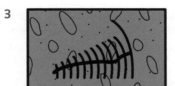

Found in only one layer in each area, so it is common but only lived for a short time.

4 That the rock layers appear in the order they were created: the younger rock layers are always on top of older rock layers, movements of the earth have not altered the order of the rock layers, the rock layers containing the same type of fossils in different locations were created at the same time.

10.1.3 Absolute dating p.207

1 **a** Carbon-14

   b Carbon-14

   c Potassium-40

   d Thorium-232

2 The amount of time required for one-half of a radioactive sample to decay forming another element

3 1408 million years (2 half lives = 2 × 704 million)

4 500 million years. Assuming rock layers are laid down at the same rate, there are 15 million years between the two aged rock layers and the fossil rock layer is roughly 1/3 of the way between the 495 mya and 510 mya rock layers.

5 **a** E (this slippage occurred after layer D was laid down), D (200 mya), A (250mya), B (layer below A so must be older than A), C (layer below B so is older than B).

   b Approximately 300mya. They are between layer C that is 350mya and layer B that is 250mya.

6

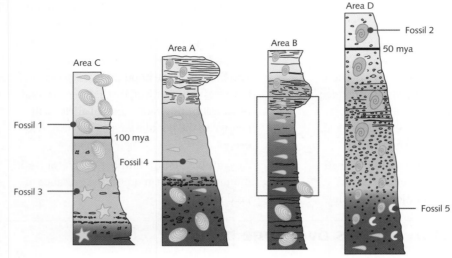

Fossil 1 is approximately 100 mya and Fossil 2 is approximately 50 mya, the rock layers they are found in have been aged. Fossil 3 is older than 100 mya, it is found in layers beneath Fossil 1. Fossil 4 is between 50 and 100 mya, found in rock layers between Fossils 1 and 2. Fossil 5 is slightly older than 50mya, it is found in same layer and layers just below Fossil 2.

10.2 Patterns in evolution p.209

10.2.1 Transitional fossils p.209

1 A fossil that shows features of both an ancestor and a descendant species. *E. watsoni* is a transitional fossil between lobe finned fish and tetrapods, it has some features of fish and some features of land tetrapods.

2

3 The depth of water in their habitat

4 Answers may vary. Organisms that could support their weight in shallow water (to enable them to move around), organisms with 'fingers' on their pectoral fins. The selection pressure of the loss of water afforded adaptations to better suit a land environment.

5 This hypothesis is refuted as fossils of *E. watsoni* show that fingers had started to evolve in aquatic fish.

6 a Both limbs contain many of the same bones, the bones have a similar arrangement.

b *Elpistostege* has an ulnare bone that is not present in the human limb, the radius and ulna are same length in human limb whereas the radius is longer than the ulna in *Elpistostege*. The humerus and digits are more elongated in the human limb.

10.2.2 Divergent evolution p.211

1 and 2

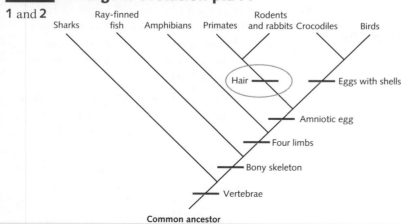

3 Vertebrae

4 Vertebrae

5 Hair, amniotic egg, four limbs, bony skeleton, vertebrae

6 Eggs with shells

7 Crocodiles and birds

8 Sharks and birds

10.2.3 Convergent evolution p.212

1 a Streamlined body shape or dorsal fin shape

b Friction/pressure of the water against their body to enable them to swim fast to catch prey.

c Sharks and dolphin share similar features but do not share a recent common ancestor, they have a very distant common ancestor, so they are not closely related. Their similar features are a result of convergent evolution as they both live in a similar environment, the ocean, and are both predators – so face similar selection pressures related to catching prey.

2 a Wings to enable flight

b Similar food source, avoid predators, need to travel long distance to feed.

c Birds and bats are not closely related as they do not share a recent common ancestor, they have a distant common ancestor. There similarities are a result of convergent evolution due to living in similar environments/ having similar selection pressures (e.g. a need to escape predators, ability to catch airborne prey).

10.3 Emergence of new species p.214

10.3.1 Allopatric speciation and Galapagos finches p.214

1 In 1976 most birds had a beak depth of between 9 mm and 10.5 mm, very few birds had beak depths between 6–8 mm and 11–13 mm. The mean beak depth is 9.5 mm.

2 A beak depth of between 9 mm and 10.5 mm is the best adapted to the available food: soft small seeds. Figure 10.11 shows that populations of finches on islands with different food sources have different sized/ shaped beaks, suggesting that there is a relationship between food type and beak shape in the finches.

3 There is minimal variation in seasonal temperatures in the different years between 1973 and 1978, with temperatures only ranging between 80–84°F in the wet seasons and 73–78°F in the dry seasons. There was significantly less rainfall between the dry season of 1976 and the end of the dry season 1977. No rain

in the dry season of 1976 and 1977 compared with 12 cm in the other two years. The wet season of 1977 only received 25 cm of rain compared to 200, 162 and 175 cm in the other three wet seasons.

4 The majority of birds have beaks between the range or 9–11 mm, with most have a beak depth between 10–11 mm, there are very few birds with beaks smaller than 9 mm. The mean beak depth is 10.1 mm.

5 The mean beak depth of the population has increased from 9.5 mm in 1976 to 10.1 mm in 1978. There are fewer birds with beak depths less than 9 cm in 1978 compared to 1976 and more birds with beak depths greater than 10 mm in 1978 compared to 1976. There is not a continuous range for beak depths under 8 mm in 1978.

6 The food available to the birds in 1978 was different to the food available in 1976, it could be larger or harder seeds that the birds are eating in 1978.

7 Natural selection: the change in food would have acted as a selection pressure on the birds. There would have been natural variation of beak sizes in the population in 1976 and the birds with larger beaks would have had an advantage over smaller beaked birds in being able to access the seeds available. Birds with larger beaks would have had a higher survival and reproductive rate than birds with smaller beaks and would have passed on alleles for larger beaks to their offspring.

8 Future generations would continue to have larger beaks unless the food source changes again, and the finch's beaks may not be suited to the food source. The larger beaked finches may become so different to the original species they may be considered a new species.

9 Allopatric speciation occurs when a population is split into different populations due to a geographical barrier. The original finch population that inhabited the Galapagos islands was split into different populations due to the distance between the islands. There was no interbreeding resulting in no gene flow between the populations on the different islands. Each island had different environments and different types of food, so different selection pressures were acting on each population of finches. Different phenotypes, especially beak sizes, had an advantage on the different islands and the advantageous alleles in each population would have been passed on to the next generation. Different mutations would have occurred in each finch population, introducing new alleles and phenotypes. Over many generations each population's gene pool could have changed enough to make them so different to each other that if brought together the different populations were no longer able to interbreed and produce fertile, viable offspring. This means that each finch population on the different islands are now considered different species.

10.3.2 Sympatric speciation and *Howea* plants p.217

» Sympatric speciation is when two species evolve from one population while living in the same geographical area.

» *Howea* plants could have gone through sympatric speciation via natural selection due to different selection pressures acting.

» The evidence shows that curly palms grow at lower altitudes and in alkaline soils compared to the kentia palms that grow at higher altitudes and in slightly acidic soils.

» By chance some of the original population may have started growing in areas that had more alkaline soil that contains less nutrients than the lower pH soils.

» The selection pressure of poorer soil would have favoured plants that could flower more quickly. These plants would have flowered before the rest of the population and only been able to reproduce with each other and not with the other *Howea* palms that have a later flowering time.

» Over time the curly palms with characteristics of early flowering on the alkaline soils would have survived and reproduced in greater numbers with passing on their alleles to their offspring.

» There would have been no interbreeding with the later flowering palms due to the time differences, so there would be no sharing of new alleles due to mutations between these two groups of palms.

» Over many generations the curly palms growing in the alkaline soil may have accumulated enough genetic differences to prevent them from being able to reproduce with the Kentia palms to produce viable and fertile offspring even if flowering time overlapped. The curly and kentia palms would now be considered separate species.

10.4 Determining the relatedness of species p.218

10.4.1 Homologous structures p.218

1

Structure	Generalised pentadactyl limb	Animal						
		Frog	Lizard	Bird	Bat	Whale	Cat	Human
Upper arm	1	1	1	1	1	1	1	1
Forearm	2	2	2	2	2	2	2	2
Wrist	9	6	7	2	3	5	6	8
Hand	19	15	20	6	7	45	18	19
Digit 1	3	3	3	1	3	5	3	3
Digit 2	4	3	4	3	1	20	4	4
Digit 3	4	3	4	2	1	14	4	4
Digit 4	4	3	5	0	1	5	5	4
Digit 5	4	3	4	0	1	1	4	4

2 **a** The upper arm and forearm have the same number of bones. Number of bones in wrist, hand and each digit can vary.

b Differences in length of different bones. The digit bones in a bat are elongated compared to other animals, the whale has a much shorter upper arm and forearm.

c Due to the function performed by the limb, e.g. whether for flying, walking or swimming. For instance, a bat does not need to have jointed digits as their only purpose is to form a support for wings.

d Differences mean that the limb is best suited to its purpose/function/type of locomotion. For example, having no joints in the digits could help make the membranes in the wings of a bat population more stable and less flexible; having shorter upper arm and forearm bones would give more strength to the flipper of the whale.

3 Similar features in these animals result from adaptive radiation, a type of divergent evolution. These organisms shared a common ancestor at some point, but as they all have occupied different ecological niches, different selection pressures have acted on them, leading to slight differences in their limbs so that they were best suited for their environment.

4 At some stage in the past these species shared a common ancestor. They have since undergone divergent evolution due to the different selection pressures present in their different environments. The cat and human are more closely related and share a more recent common ancestor due to having the least numbers of differences in the number of bones in each limb part.

10.5 Molecular evidence for relatedness between species p.221

10.5.1 The concept of molecular homology p.221

1 Yeast, wheat, Drosophila, dogfish, chicken, pig, chimpanzee, human

2 Amino acid positions 1, 2, 6, 17, 18, 19, 20 are crucial as at these positions the amino acid is the same in all organisms, suggesting the presence of one particular amino acid (usually a conserved region) at each of these positions is needed for cytochrome C to function correctly. Positions 3, 9 and 22 all have 4 different amino acids in each of these positions, suggesting the amino acid at these positions is less important to the correct functioning of cytochrome C.

3 Yes, the evidence supports how these organisms are classified. Yeast and wheat are classified into different kingdoms, fungi and plants, compared to the rest of the organisms. You would expect them to have a more distant common ancestor with the animals.

Drosophila is an invertebrate, so it would be more distantly related to the vertebrates than they are to each other.

10.5.2 Assembling phylogenetic trees p.222

1

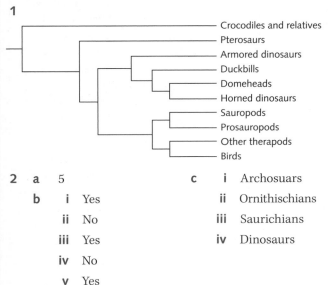

Crocodiles and relatives
Pterosaurs
Armored dinosaurs
Duckbills
Domeheads
Horned dinosaurs
Sauropods
Prosauropods
Other therapods
Birds

2 a 5
  b i Yes
    ii No
    iii Yes
    iv No
    v Yes
  c i Archosuars
    ii Ornithischians
    iii Saurichians
    iv Dinosaurs

10.6 Chapter review p.224

10.6.1 Key terms p.224

1a Radioactive dating	a method for determining the age of a rock or fossil based on the predictable rates of decay of naturally occurring radioactive isotopes present
b Absolute dating	the process of determining the age in years of rocks and the fossils they contain on the basis of the physical or chemical properties of materials in the rock
c Reproductive isolation	sexual reproduction can no longer occur freely among any adult members of the population
d Speciation	the evolution of one or more new species from an ancestral species
e Allopatric speciation	speciation that occurs when members of an ancestral population become geographically separated and each isolated population evolves into a new species
f Sympatric speciation	when two species evolve from an ancestral population while still inhabiting the same geographical area
g Divergent evolution	when members of the population develop adaptations to the different selection pressures over many successive generations, they eventually become new species
h Adaptive radiation	a single species diversifies relatively rapidly into many new species because of the availability of many different ecological niches
i Convergent evolution	when organisms that are not closely related, independently evolve similar traits as a result of having to adapt to similar environments or ecological niches
j Vestigial structure	structures found in organisms that have lost most, if not all, of their original function in the course of evolution; in ancestral organisms the structures served a purpose, but in their descendants the structures become atrophied or rudimentary
k Homologous structure	anatomical features of different organisms that have the same basic underlying structure but different functions
l Analogous structure	anatomical or morphological features of different organisms that have the same function but not the same basic underlying structure
m Molecular homology	the similarity of patterns in the nucleotide sequences of DNA or amino acid sequences of polypeptides as evidence for a common evolutionary origin
n Phylogenetic tree	a branching diagram showing the evolutionary relationships between species; groups joined together in the tree are believed to have descended from a common ancestor
o Cladogram	a phylogenetic tree that depicts a hypothesis about the evolution of a group of organisms from a common ancestor

2 Answers will vary. Check definitions against the textbook glossary.

1 B

2 A

3 B

4 D

5 B

6 **a** Convergent evolution (1). The three organisms are not closely related but share similar features due to being exposed to a similar selection pressure in their aquatic environment (1).

b **i** One of the following: climate change, disease, lack of genetic diversity, environmental change leading to loss of food source, new predator (1)

ii Four of the following: after death it was quickly covered with sediments, body is not eaten by scavengers, there is a lack of oxygen to prevent decay, over time covered with layers of sediment that form sedimentary rocks, erosion or shifting of rock layers expose the fossil in the rock layers (4)

Chapter 11 Human change over time p.229

Remember p.229

1 Allopatric speciation occurs when members of one species are separated by a geographic barrier and each population evolves into a new species. Sympatric speciation when two species evolve from an ancestral species while occupying the same geographic area.

2 No, the fossil record is biased towards organisms that fossilise readily.

3 A fossil is preserved remains or traces of an organism. It can give some information about organisms that have died and become extinct.

4 Carbon-14 is an unstable isotope and over time it decays to become nitrogen-14. This happens at a known rate and absolute age of a fossil can be found by determining the amount of carbon-14 left in the fossil.

5 Endothermic, body hair, unique exocrine glands: mammary, sweat, sebaceous, scent, four chambered heart, diaphragm, three bones in inner ear, single jawbone, have three types of teeth: incisors, canines, and molars

6 Horse and cow – they are both mammals, the other two pairs each contain a mammal and an animal of another taxonomic class.

7 The evolutionary relationship between groups of organisms, a node represents the common ancestor of the two organisms that diverge from the node

8 The length of the branches in a phylogram represent the number of nucleotide or amino acid changes that occurred during the evolution of each species, a cladogram does not show this.

11.1 Taxonomy of modern humans p.230

11.1.1 Taxonomy p.230

1

Level of taxonomy	Humans	Features that all members share	Other animals in this group
Kingdom	Animalia	Multicellular, eukaryotic cells, no cell walls, heterotrophic	Lemur, ape, mandrill, orangutan, chimpanzee
Phylum	Chordate	At some stage in their life possess a notochord, a dorsal nerve cord, pharyngeal slits, and a post-anal tail	Lemur, ape, mandrill, orangutan, chimpanzee
Class	Mammalia	Endothermic, body hair, unique exocrine glands: mammary, sweat, sebaceous, scent, 4 chambered heart, diaphragm, three bones in inner ear, single jawbone, have 3 types of teeth: incisors, canines, and molars	Lemur, ape, mandrill, orangutan, chimpanzee
Order	Primates	Hands and feet with five digits, opposable digit, nails instead of claws, forward-facing eyes enabling stereoscopic colour vision, large cranium for body size, flexible spine, able to rotate through hips and shoulders	Lemur, ape, mandrill, orangutan, chimpanzee
Family	Hominidae	No tail, gliding wrist joint, dentition includes eight premolar teeth, molars with five cusps and grooves in a Y-shaped pattern, broad flattened rib cage, arms generally longer than legs (except for humans)	Ape, orangutan, chimpanzee

(continued)

Level of taxonomy	Humans	Features that all members share	Other animals in this group
Genus	Homo	Bipedal mode of locomotion	
Species	sapiens	Relatively hairless, foramen magnum located close to the centre of the base of the skull, Small face with projecting nose bone, rounded back of the skull, s shaped spine, shallow bowl shaped pelvis, arched foot, big toe in line with other toes, femur angled inwards, precision grip, large cranium for size, expanded prefrontal cortex, small brow ridges, parabolic jaw, reduced canines	

2 a   Suborder

   b   Parvorder

   c   Tribe/Genus

   d   *Pithecliidae, Cebidae, Atelidae, Hylobatidae, Hominidae*

   e   Chimpanzees

   f   Chimpanzees and bonobos are more closely related than gorillas and orangutans. Chimpanzees and bonobos are classified in the same genus whereas the lowest classification level that both gorillas and orangutans have in common is family. Genus is lower down the classification hierarchy and members of the same genus are more closely related than members on the same family.

   g   They are primates with no tail, gliding wrist joints, eight premolar teeth, molars with five cusps and a Y-shaped pattern, broad flattened rib cage, arms generally longer than legs, opposable thumbs/toes. Large brain for body size, five digits, flat nails instead of claws, colour vision, forward facing eyes, rotatable hip and shoulder sockets, flexible spine.

   h   Bipedal locomotion, foramen magnum centred under base of skull, enlarged cerebral cortex, less body hair, small face with projecting nose bone, rounded back of the skull, S-shaped spine, shallow bowl shaped pelvis, arched foot, big toe in line with other toes, femur angled inwards, precision grip, small brow ridges, parabolic jaw, reduced canines, legs longer than arms, no opposable toe.

3 a   30 mya

   b   approx. 7 mya

11.2 Adaptations that define humans p. 233

1

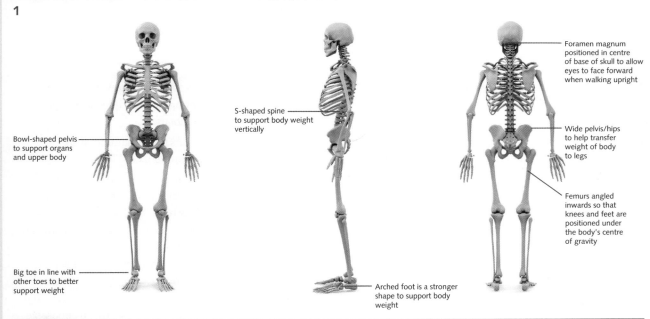

Bowl-shaped pelvis to support organs and upper body

Big toe in line with other toes to better support weight

S-shaped spine to support body weight vertically

Arched foot is a stronger shape to support body weight

Foramen magnum positioned in centre of base of skull to allow eyes to face forward when walking upright

Wide pelvis/hips to help transfer weight of body to legs

Femurs angled inwards so that knees and feet are positioned under the body's centre of gravity

11.2.2 Evolution of two-legged walking p.233

1   Evidence. It uses observations and evidence from fossil finds of named ape species to support statements. However, there are no details of the author or references to support statements.

2   That bipedalism only evolved in humans (hominins).

3   If scientists are correct in analysing the evidence presented by *D. guggenmosi*, then bipedalism evolved millions of years before the appearance of humans.

4   Skeletal evidence found in fossils of apes that lived 11.6 mya: knee and ankle structures that suggest weight bearing adaptations, vertebrae that suggest their backs were long and flexible, pelvic adaptations that suggest an upright stance.

5

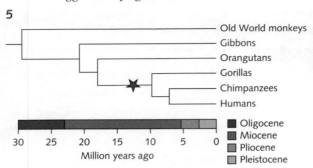

6   **a**   Limited evidence: 'the available remains of the fossilised spine are fragmentary and deformed.'

b   More and better-preserved fossil evidence of *D. guggenmosi* structures that would indicate bipedal stance, e.g. spine, ankles, pelvis, leg and knee.

7   The accepted evolutionary timeline of humans and apes would have to be reconsidered. Did the mode of locomotion used by apes and chimps evolve from a bipedal ancestor?

11.2.3 **Human ingenuity p.235**

1 and **2** Writing in question 2 should be messier and less legible than in question 1.

3   It is more difficult/impossible to write neatly and legibly without use of an opposable thumb.

Expansion of the cranium p.236

4

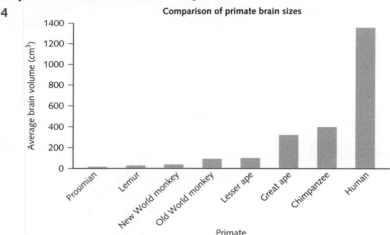

5   Prosimian: 8, Lemur: 7, New World monkeys: 6, Old World monkeys: 5, Lesser ape: 4, Great ape: 3, Chimpanzee: 2

6   As the evolutionary distance from humans decreases, brain volume (cm³) in primates increases, there is an inverse exponential relationship between evolutionary distance from humans and primate brain volume (cm³).

7   Increased cognitive ability: ability to learn, plan, evaluate, make decisions, apply new knowledge, develop complex communication methods, abstract thinking and problem solving. Allowed for technological development: development of tools and cultural development: forming complex social groupings

8   Cranial capacity: human skull is about three times the size of the chimpanzee skull. Human skull has a rounded braincase.

Prognathism: human skulls are relatively flat and chimpanzee skulls protude beyond the top of the skull.

Dentition: human teeth are small and regular in size with a smaller jaw; chimpanzees have sharp, large canines and large premolars.

Brow ridges: in humans they are minimal, in chimpanzees they are pronounced.

9   Answers will vary. Humans are bipedal and relatively hairless whereas other primates walk on all four limbs and are covered in body hair. To allow bipedalism in humans the thigh bone is angled inwards, they have an S shaped spine and a shallower more bowl like pelvis compared to other primates which have long narrow pelvis and a forward leaning C shaped spine. Human feet have big toes in line with the other toes, transverse and longitudinal arches, other primates have an opposable big toes and only longitudinal foot arches. Human skulls have a larger cranium and a much larger frontal lobe, less protruding jaws and brow ridges compared to other primates. The jaw of humans is parabolic whereas the jaw of other primates is rectangular. Human canines are smaller than other primates.

11.3 Meet the ancestors p.238

11.3.1 From *Australopithecus* to *Homo sapiens* p.239

1

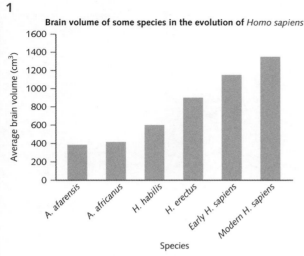

Brain volume of some species in the evolution of *Homo sapiens*

2 Cross out *Australopithecus africanus*

3 Circle *Homo habilis*

4
a Average brain volume of 400 cm³ puts *H. floresiensis* between *A. farensis* and *A. africanus*

b Skull: prominent brow ridges, rectangular lower jaw, foramen magnum at rear of skull, protruding jaw, smaller cranium – would all be more ape like characteristics. Small brow ridges, parabolic lower jaw, foramen more centered under the skull, flatter face – would all mean closer to *H. Sapiens*.

Pelvis – more bowl shaped – closer to *H. sapiens*. Longer and less bowl shaped – more ape like. Arm and leg length – longer arms compared to legs – more ape like. Same size arms or slightly shorter arms compared to legs – closer to *H. sapiens*

c Allopatric speciation: the original populations were geographically isolated on an island with no gene flow with other populations. Over time new mutations would have appeared in this population and due to selection pressures on the island the population would have evolved into a new species.

d Answers may vary. The arrival of other *Homo* species to its island home created competition for resources and *Homo sapiens* out competed *H. floresiensis* for resources, causing them to become extinct by other *Homo* species when they arrived on their island. Low genetic diversity meant they could not survive a change in their environment

e If only one fossil was found, it could be concluded that the difference in features are due to some form of disease or dwarfism. If it were another species if would be expected that there would be more than one fossil found.

f Answers may vary. Further fossil evidence that would allow a comparison to other hominin species, size of brow ridges, lower jaw shape, size of teeth, flatness of face can all be used to determine evolutionary relationships of hominins. Comparing other skeletal features to those of other hominins to determine shared features could also help position *H. floreseiensis* on the evolutionary tree. Extracting and analysing DNA samples from fossils would allow comparisons to other available DNA sequences from hominins.

11.4 Modern humans and Neanderthal p.241

11.4.1 Evolution of modern humans p.241

1–4

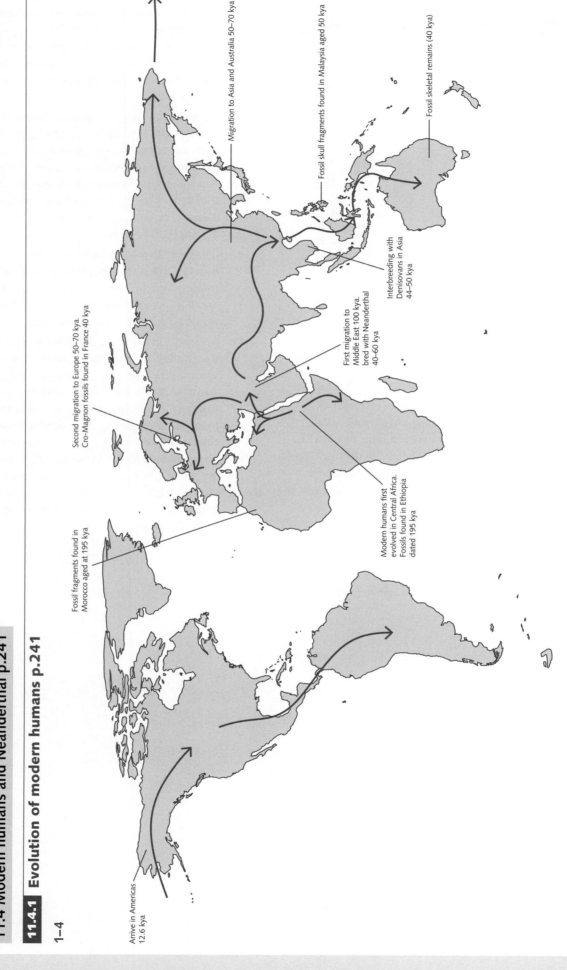

Arrive in Americas 12.6 kya

Fossil fragments found in Morocco aged at 195 kya

Second migration to Europe 50–70 kya. Cro-Magnon fossils found in France 40 kya

Migration to Asia and Australia 50–70 kya

Fossil skull fragments found in Malaysia aged 50 kya

Fossil skeletal remains (40 kya)

First migration to Middle East 100 kya. bred with Neanderthal 40–60 kya

Interbreeding with Denisovans in Asia 44–50 kya

Modern humans first evolved in Central Africa. Fossils found in Ethiopia dated 195 kya

11.4.2 Australian settlement p.243

1 mtDNA is inherited from the mother only and it has a known mutation rate, providing a much more direct genetic lineage to ancestral populations. Nuclear DNA is a combination of both paternal and maternal DNA and is subject to recombination which makes it more difficult to trace the DNA sequences back to ancestral populations.

2 The population migrated from New Guinea which is at the top of Australia, along the east and west coasts because the environmental conditions on the coast provided more resources to support populations, the arid centre of Australia would have been a barrier to migration.

3 O, R and M

4 P, S and M

5 The dry and arid centre of Australia

6 P

7 Only 111 hair samples from the South Australian museum were analysed. This is a very small sample and would not have been representative of all the different indigenous populations.

8 The age of fossils is evidence of the presence of Aboriginal and Torres Strait Islander communities in different locations for over 40 000 years in some places up to 60 000 years. Evidence from haplogroups suggests that these populations stayed in the same location, showing an ability to make the most of available resources. Fossil artefacts such as cave paintings show that knowledge of their land and Country was passed onto the next generation. Prolonged habitation of a location and passing on shared knowledge of the location demonstrates connection to Country and Place.

9 European colonisation removed Aboriginal and Torres Strait Islander peoples from their traditional homelands and Country and prevented them from practicing and sharing culture that connected them to their Country and stopped the passing on of traditional knowledge about their Country.

10 Answers will vary. The first extract is evidence based, using evidence from DNA testing and fossil and artifact finds to support statements. The other two extracts are anecdotal, or a narrative presented by one person. The information presented supports accepted understanding of the content presented.

11.4.3 Relationship between modern humans and Neanderthals p.246

1 Suggests that Neanderthals and modern humans lived in the same area and interacted with each other for 5000 years longer than originally thought.

2 A molar and bone fragments identified as from modern humans were found amongst Neanderthal artefacts, tools, beads, and pendants.

3 Answers will vary, must be a prediction or explanation. Neanderthals became extinct because they were outcompeted by modern humans. Neanderthals became extinct because they lacked genetic diversity and could not survive changes to their environment.

4 Answers will vary depending on the hypothesis. Outcompeted: evidence that Neanderthals become extinct soon after the arrival of modern humans. Lack of genetic diversity: compare sequences of Neanderthal DNA samples to measure amount of similarity.

5 Modern humans bred with Neanderthals after they migrated out of Africa and produced fertile, viable offspring that carried some Neanderthal DNA. Members of these populations created European and Asian populations passing on the Neanderthal DNA. The population that remained in Africa did not meet Neanderthal populations and did not interbreed with them.

11.5 Chapter review p.247

11.5.1 Key terms p.247

1 *Homo floresiensis, Homo sapiens, Homo neanderthalensis, Denisovan*

Homo erectus

Homo habilis

Homo

Australopithecus afarensis

Australopithecus

Danuvius guggenmosi

2 a Cranial capacity is the volume of the brain case whereas cognitive capacity is related to innate intelligence

b Haplotype is a unique combination of genetic mutations in the mtDNA or the Y chromosome in a group of organisms that carry the same mutation and indicates common ancestry when shared by different organisms. Haplogroup is a group of organisms with similar haplotype, indicating shared ancestry.

c A hominoid is a tailless primate, an ape. A hominin is a hominoid that is bipedal, walks on two legs.

d An opposable thumb is a digit that can be put against each of the other digits of a hand, this allows the digits to grasp and hold objects. This allows precision grip, the bringing together the tip of the thumb and other digits for precise manipulation of objects.

e Locomotion can be bipedal, walking on two legs, or quadrupedal, using all four limbs to walk.

f A sagittal crest is a raised ridge of bone running along the midline of the top of the skull whereas a sagittal keel is a thickening of bone along the midline of the skull.

g Oviparous animals lay eggs in which the embryo develop and offspring hatch. Viviparous animals give birth to live young, the embryonic development occurs within the mother's body.

11.5.2 **Exam practice p.249**

1 C

2 D

3 A

4 D

5 C

6 a Fossils of arm and leg bones showing that arms were longer than legs

b One of *H. sapiens*, Neanderthals, Denisovan and *H. floresiensis*

c Any two of: foramen magnum centred under the skull, angled femur bones, shallow bowl-shaped pelvis, S shaped spine, big toe in line with other toes, arch in foot

d Made and used tools